6/08

W9-AMC-487

Love and Sex with Robots

LOVE + SEX

— WITH —

ROBOTS

THE EVOLUTION OF HUMAN-ROBOT RELATIONSHIPS

DAVID LEVY

HARPER

An Imprint of HarperCollins*Publishers*

www.harpercollins.com

HarperCollins books may be purchased for educational, business, or sales promotional use. For information, please write: Special Markets Department, HarperCollins Publishers, 10 East 53rd Street, New York, NY 10022.

FIRST EDITION

Designed by Fritz Metsch

Library of Congress Cataloging-in-Publication Data

Levy, David.
 Love and sex with robots : the evolution of human-robot relations / David Levy.
 Includes bibliographical references.
 ISBN: 978-0-06-135975-0
 1. Human-computer interaction.   I. Title.
 QA76.9.H85L48 2007
 629.8'92—dc22   2007027508

07 08 09 10 11 NMSG/RRD 10 9 8 7 6 5 4 3 2 1

To "Anthony,"* an MIT student who tried having girlfriends but found that he preferred relationships with computers. And to all the other "Anthonys" past, present, and future, of both sexes.

*Described by Sherry Turkle in *The Second Self*.

Acknowledgments

I would like to thank those who assisted in various ways during the writing and editing of this book.

Christine Fox and Bill Yeager carefully read what I had expected to be a near-final draft, and each of them contributed a number of important suggestions for improvements. Ray Kirsch and Heather Allan read a subsequent draft and provided encouragement. My literary agent, Mollie Glick, and my editor at HarperCollins, Rakesh Satyal, made many further suggestions that enhanced the typescript. Rakesh also gave me a few useful lessons that improved my knowledge of the American language.

Dr. Cynde Moya offered some helpful advice on early-twentieth-century sex artifacts and brought to my attention two of the exhibits from Magnus Hirschfeld's former collection (the artificial vagina on page 178 and the fornicatory dolls on page 181). She also kindly provided me with images of both of these, as reproduced in her Ph.D. thesis "Artificial Vaginas and Sex Dolls: An Erotological Investigation." Alan Pate provided the original source information for explaining the origin of the term "Dutch wives." Professor Jaap van den Herik offered encouragement to the point of inviting me to submit an academically redrafted version of my text for a Ph.D. at the University of Maastricht. Andrew Keatley and Alastair Levy assisted with the technicalities of some of the images. Kimballe Thomerson was my Japanese-speaking intermediary in my communications with the manufacturer Orient Doll.

I also wish to thank the following for granting me permission to reproduce certain images and text: Peter Menzel/Science Photo Library for Peter Menzel's photograph of Kismet (page 13); the *New Yorker* for Peter Steiner's cartoon "On the Internet, nobody knows you're a dog." (page 42); Getty Images for the photograph of the Repliee Q1 robot (page 146); Chris Buck for his photograph of David Hanson's robot head (page 163); John Suler for the extract from his article "Mom, Dad, Computer (Transference Reactions to Computers)" in his electronic book *Psychology of Cyberspace* (page 191); Jane Treays and the BBC for the quotations from the TV documentary *What Sort of Gentleman Are You After?* (pages 214–15); Orient Industries for the company's photograph of one of their Orient Doll products (page 248); *Asian Sex Gazette* (www.asiansexgazette.com) for Mark Schreiber's article "A Jewel in Japan's Hi-Tech Crown: Sex Dolls" (pages 249–50); Mainichi newspapers for Ryann Connell's article "Rent-a-Doll Blows Hooker Market Wide Open" (pages 251–52); Dave Lampert for the image of the Sybian "Lovemaster" sex machine (page 254) and its inserts (page 255); and Paul Gaertner for the photograph of the Stallion XL sex machine (page 258).

According to the United Nations Economic Commission for Europe's World
Robotics Survey, in 2002 the number of domestic and service robots more
than tripled, nearly outstripping their industrial counterparts. By the end
of 2003, there were more than 600,000 robot vacuum cleaners and lawn
mowers, a figure predicted to rise to more than 4 million by the end of next
year. Japanese industrial firms are racing to build humanoid robots to act
as domestic helpers for the elderly, and South Korea has set a goal that
100 percent of households should have domestic robots by 2020.
"Probably the area of robotics that is likely to prove most controversial is
the development of robotic sex toys," says Dr. Christensen. "People are
going to be having sex with robots in the next five years," he says. "Initially
these robots will be pretty basic, but that is unlikely to put people off,"
he says. "People are willing to have sex with inflatable dolls, so initially
anything that moves will be an improvement."

—*The Economist*, June 8, 2006, quoting Henrik Christensen, chairman of the
    European Robotics Network at the Swedish Royal Institute of Technology in
    Stockholm

Contents

Love and Sex with Robots

Introduction

Recent research shows that people perceive and treat robots not just as machines, but also as their companions or artificial partners.

—Alexander Libin and Elena Libin, 2004[1]

At the dawn of the twenty-first century, mankind is experiencing an era of phenomenal scientific and technological achievement. Whole disciplines of science that were unheard of even a few decades ago are now making possible amazing feats in areas such as cell-phone technology, computer technology, space research, and medicine. Furthermore, our scientific knowledge is growing at a rate that is itself increasing. The more we know about a science, the more quickly we may use our knowledge to discover even more within that science. This has been very much the case in the field of computing, a science that (like me) was in its infancy in the early 1950s. In those days each of the few computers that *had* been built would fill a room and cost a fortune. And although articles about computers appeared from time to time in the popular press, few people had any idea what these newfangled machines could be used for. When, in 1943, an American company called International Business Machines first considered the possibility of manufacturing computers on a commercial basis, the company's founder and president, Thomas J. Watson, pessimistically predicted, "I think there is a world market for maybe five computers." How wrong he was! Instead of the computer's being something of a commercial white elephant, it became the product for which IBM is best known. And by 1981 the computer had become so ubiquitous in industry, in the office, and in academic life that IBM launched a whole new product category called the personal computer, the PC, a computer that was not only more powerful than the multimillion-dollar machines of twenty years earlier but was also affordable for many families and individuals.

Commensurately with this dramatic growth in the popularity of the computer as a tool for all to use, computer science became a subject that was increasingly studied at universities and research institutes. And within computer science there came an even newer discipline, called artificial intelligence,* the science of making computers that can think. Every science has its own divisions and subdivisions, and artificial intelligence (AI) is no exception. Developing programs to play games such as chess falls within the boundaries of a division of AI called "heuristic programming."† Programs that carry on conversations or translate from one language into another are encompassed within the AI discipline of "natural language processing." And among the other disciplines within AI there is robotics.

The word "robot" was suggested by Josef Čapek‡ in discussion with his more famous brother, the Czechoslovak writer Karel Čapek. It is derived from the Czech *robota* (forced labor) and was first revealed in the West when Karel used it in the title of his play *Rossum's Universal Robots* (*R.U.R.*), an immediate hit when it was first shown on Broadway. The literal meaning of "robot" is "worker." The robots in Čapek's play were creature machines, resembling humans in appearance, designed and built to serve as workers for their human masters.

Although the word "robot" was new in the early 1920s, the idea of an artificial form of life was by no means a new one in Čapek's day. Inventors and engineers had for millennia devised automata that simulated some of the functions of living creatures. One of the earliest to do so was Heron§ of Alexandria, who lived in the first century A.D.** Among many inventions that were mechanical marvels for their time, Heron

---

*Also called machine intelligence.
†Heuristics are commonsense but often imperfect rules of thumb, designed to speed up the process of finding solutions to certain types of problems.
‡The accented letter Č is pronounced like *ch* in "chicken."
§Also known as Hero.
**Until 1938 there had been some doubt about Heron's dates, some sources believing him to have lived around 150 B.C. and others around A.D. 250. Then Otto

constructed some water-powered mechanical birds, entire flocks of them, that even emitted realistic chirping sounds created by a water-driven device.

The public's fascination for automata reached its first peak in France in the eighteenth century. One example of this genre was a menacing mechanical owl set amid a group of smaller birds, designed in 1644 by the French engineer Isaac de Caus. The smaller birds would flutter their wings and chirp while the owl slowly moved on a pivot to face them. As the owl's face turned toward the smaller birds, appearing to threaten them, they became still and stopped their chirping. When the owl's face then turned away from the group, the smaller birds came alive again. The whole mechanism was driven by a water wheel that controlled the actions of each bird by means of a metal cylinder, the surface of which was embedded with pins, just like a music box. As the cylinder turned via the force of the water, the pins on the cylinder would engage with a music box–like mechanism so that each pin created its own effect or movement in one of the birds.

Following de Caus's example, at least two other French automaton inventors also used birds as the embodiments for some of their mechanical marvels. In 1733 an inventor named Maillard designed a mechanical swan that would paddle through the water while its head moved slowly from side to side. Maillard's idea was as simple as it was clever: a paddle wheel, similar to those found in the Mississippi River steamboats, propelled the swan forward while simultaneously connecting, via a system of gears, with the swan's head; as the paddle wheel rotated, it thus served a dual purpose, creating the forward motion of the swan's body and the simultaneous side-to-side motion of its head. An even more advanced idea, and a more entertaining example of this genre, was a mechanical defecating duck, the creation of Jacques Vaucanson. The duck could bend its neck, move its wings and its feet, and it could "eat." It would stretch out its neck to peck at corn offered

---

Neugebauer noted that Heron had written about a "recent eclipse," which, from the information given by Heron in his writings, was dated to one that took place at Alexandria on March 13, A.D. 62.

by a human hand, then swallow, digest, and finally excrete it, the corn having been turned into excrement by a chemical process, according to Vaucanson. In fact the "digestion" and "excretion" processes were parts of a hoax. The corn, once eaten, was held in a receptacle at the lower end of the duck's throat, while the duck's "excrement" was not genuine duck droppings but some other material that had been inserted in the duck's rear end prior to the demonstration.

That Vaucanson's duck did not actually digest its food and defecate in no way diminishes its contribution as a precursor to humanlike robotics. One of the principal achievements of Vaucanson and his peers was the stimulation of widespread interest in the mechanical aspects of what is now known as artificial life. That era saw the creation of automata that could not only eat but also breathe; automata with soft skin, flexible lips, and delicately moving jointed fingers.* A remarkable example of a humanoid automaton was a birthing machine designed in the mid-eighteenth century by Angélique du Coudray, mid-wife to the royal court of France. The purpose of this machine was to assist in the teaching of midwifery, as a result of which many examples of the machine were made and sent to doctors and midwives throughout France. Du Coudray's machine was made of wicker, stuffed linen and leather, dyed in various flesh-tone colors, some pale and some of a deeper red, to simulate the softness and appearance of a woman's skin and organs. The pelvic bones of human skeletons were used in some of her machines, and sponges soaked in liquids colored red and other hues were used inside the machine, releasing their simulated bodily fluids at appropriate stages of the lectures on the birthing process.†

While Vaucanson and his peers managed the simulation of physiological and other natural bodily processes, there were other inventors who focused on simulating the processes of thought. One of the

---

*An extremely comprehensive and valuable account of the history of such automata is provided in two papers by Jessica Riskin.[2, 3]
†The only known example still extant is in the Musée Flaubert in Rouen, France, a museum of the history of medicine. Two photographs appear in Nina Gelbart's *The King's Midwife*.

best known of these peers was, like Vaucanson, also famous for a machine that turned out to be a hoax. In the closing years of the eighteenth century, Baron Wolfgang von Kempelen, a scientific adviser to the royal court of Vienna, designed a chess-playing automaton in the guise of a Turk seated on a wooden box. Despite Kempelen's assurances to the contrary, and his magician-like demonstrations to convince his audiences that the wooden box contained nothing untoward, there was in fact a (small) strong human player secreted in the box, a player who vanquished all chess enthusiasts who tried their luck against "the Turk."

On the far side of the world, the Japanese interest in robotics also dates back to the eighteenth century, during the Edo period in Japanese history, with the design of a tea-carrying doll, called *karakuri*. When a host, seated opposite his guest, placed a cup of tea in the doll's hands, it carried the cup to the guest, who then took the cup from the doll, whereupon the doll stopped moving. After drinking the tea, the guest put the cup back into the doll's hands, the weight of the cup causing the doll to spin around and return to the host with the empty cup. These dolls were fashioned in the form of a child, with the technology hidden inside, creating an aura of mystery and magic. Rather than designing their automata to look like animals, as many of the French inventors had done, the Japanese had realized more than two hundred years ago that automata are more appealing if presented in the guise of humans, a realization that anticipated some of the research described here in chapter 4.

These eighteenth-century marvels did much to create a climate of interest in the notion that human and animal bodily and mental processes can be successfully simulated. By 1830, walking dolls were being constructed and exhibited in Paris, and soon thereafter came dolls with moving eyes. Next came dolls that could eat, drink, dance, breathe, and swim (in three different strokes: backstroke, breast stroke, and crawl). And for those that could drink, one inventor, Leon Bru, created an artificial bladder, so that after taking a drink his dolls could pee. It was in this climate, and with the benefit of the nineteenth-century development of electricity, that the idea of robots as we now see them began to take root.

Karel Čapek's vision was of robots that could think for themselves, robots with feelings, robots that could fall in love with each other. In *Rossum's Universal Robots,* one of the scientists at the robot factory came up with the idea of endowing the robots with emotions, which led to their developing feelings of resentment about being treated like the slaves of human masters. Čapek had the foresight to predict what some people today fear about a future with robots—that they will "take over the world"—and in his play, the robots decided to rebel and kill all human beings.

When it premiered in New York in 1922, *Rossum's Universal Robots* was hailed by one critic as a "brilliant satire on our mechanized society," and the concept of robots as Čapek envisioned them was taken up by several science-fiction writers, most notably Isaac Asimov. In 1940, Asimov reacted to the plethora of books and stories that had already been published in which man created robots that became killers. Asimov proposed three "laws of robotics," later augmented by a fourth law, all designed to safeguard mankind's interests in the face of whatever ideas the robots of the future might develop.*

Since the birth of the science of artificial intelligence in the mid-1950s, gigantic strides have been made in the quest for a truly intelligent artificial entity. The defeat of the world's best chess player, Garry Kasparov, was just one of these strides. Others include the creation of computer programs that can compose music that sounds like Mozart or Chopin or Scott Joplin, at the operator's behest; programs that can draw and paint better than many human artists whose work today hangs in art galleries and in the homes of wealthy collectors; and programs that can trawl the Internet and write news stories based on the

---

*Asimov's laws, the first three of which were introduced to the public in his 1942 short story "Runaround," are "1. A robot may not injure a human being or, through inaction, allow a human being to come to harm. 2. A robot must obey the orders given it by human beings except where such orders would conflict with the First Law. 3. A robot must protect its own existence as long as such protection does not conflict with the First or Second Law." Later Asimov added the Zeroth Law: "A robot may not injure humanity, or, by inaction, allow humanity to come to harm."

information they gather, stories written in a style of which most journalists would be proud. Then there are expert systems—programs that incorporate human expertise to enable them to solve analytical problems normally assigned to human experts. Such programs are powerful tools for medical diagnosis, and they have also proved to be highly competent in a wide diversity of other fields, such as prospecting for minerals, making political judgments, detecting fraudulent uses of credit cards, and making recommendations in court cases to judges and lawyers, even advising defendants how to plead. These are not examples of what might be in the future—they are just some of the accomplishments of AI in its first fifty years. During the second half of the twentieth century, science fiction became a hugely popular literary form, paralleling the development of the science of artificial intelligence. One exemplar of this parallel is the computer Hal in Arthur C. Clarke's *2001: A Space Odyssey.* Hal crushes David, the human hero, at chess, mirroring the defeat of Garry Kasparov, four years prior to 2001, by IBM's Deep Blue chess-playing computer.

It was industry that prompted the initial Japanese research into robotics. And although it was also in industry that robots were first employed to replace humans (just think of car factories), the major thrust of robotics in Japan during the 1990s and into the first few years of the present century has been in "service" robots. At first, service robots were mainly used for drudgery-related tasks—cleaning robots, sewer robots, demolition robots, mail-cart robots, and robots for a host of other tasks, such as firefighting, refueling cars at gas stations, and in agriculture. But after the service-robot industry became well established in Japan, the country's robot scientists turned their attentions to the realm of personal robots, to be used at home by the individual. Mowing the lawn and vacuuming the carpet have both become tasks that in a slowly but steadily increasing number of homes are now undertaken by robots. Similarly, robots are beginning to be used in education, and Toyota has announced that by 2010 the company plans to start selling robots that can help to look after the elderly and to serve tea to guests in the home. This trend, from the use of robots in industry to their use in service tasks and now in the home, represents a shift toward an increasing level of interaction between robots and humans. In industry a button is pressed and the robot springs into action on the assembly line, working away on a repetitive task with little or no need for supervision until the daily quota of cars or whatever has been manufactured. If a robot can manage assembly-line tasks once, it can manage them time and again. If your car works well when you buy it, you can reasonably assume that the next guy's car will also work well, and the next, and so on. That is the great advantage of industrial robots—not only do they do the job as often as is needed, they do it as well the hundredth time, and the thousandth, as they did the first time. And it is this advantage of repetitive excellence that makes the industrial robot so impersonal.

A service robot does not normally need to perform its designated task time and again, one immediately after another. Instead it is there like a butler, to be at the beck and call of the individual when needed to mow the lawn or vacuum the floor, a task that might occur only once a day, once a week, or even less often. But to use a service robot requires of its owner much more interaction than with industrial robots. The owner often needs to collaborate with the robot—by bringing it onto the lawn, for example—before the robot can start work, and then to wheel the robot away again when its task has been completed. Not always, however. Some lawn-mower robots take themselves off to the garden shed when it rains or when their work is done, and some even recharge themselves by wandering over to the power socket and connecting themselves when their batteries are low—an electronic parallel of "I'm hungry, Mommy, so I'm going to take some food from the fridge."

As with many other lines of research in robotics, the first fully working androids (human-shaped robots) were developed in Japan. Development on androids started at Waseda University in the 1970s, many years before the states of the art in computing, vision technology, and various branches within artificial intelligence reached the levels needed in a twenty-first-century autonomous android. The 1980s saw a burst of engineering effort in artificial hands and other limbs, but at the time there were very few industrial applications for such technologies, and so the momentum from those efforts was not sustained throughout the 1990s. But after a gap of a decade or so, Waseda University and other Japanese robotics groups are now making good use of that earlier research-and-development effort.

The initial forays by roboticists into the world of fully interactive autonomous robots focused on entertainment, with creations such as robot toys, robot pets, and robots that play sports. Simple electronic cats and dogs have been shown to provide psychological enrichment for humans, being both pleasurable and relaxing to play with. More recent research has started a trend for interactive robots that act as human helpers, showing visitors around museums, caring for hospital patients and the elderly, and providing therapy to cope with emotional problems. Japanese researchers have shown, for example, that the mood of a child can be improved by interaction with a robot and that robots are able to encourage problem children to communicate more with each other and with their caregivers.

Toys such as Furby, Tamagotchi, and Robosapien and the virtual characters that inhabit the worlds created by computer-game designers are part of an evolutionary technological process that has turned the simulation of cognizance and perception into something much more—a force with massive potential. This force already manifests itself as an expectation, by many people, that the toys and computer programs with which they interact today will exhibit signs of life. We know that it is an artificial form of life, but the expectation that something will exhibit even an artificial form of life is a significant step toward the acceptance of such forms as real. And that day is not as far into the future as you might believe. To some extent at least, the acceptance of robots as entities capable of interesting, useful, and rewarding interaction with humans has already arrived. In Chevy Chase, Maryland, a nonprofit organization called the Institute of Robotic Psychology and Robotherapy has been set up to study some of the fundamental questions of mind, emotion, and behavior that relate to human-robot interaction. Robotic psychology focuses on human-robot compatibility, while robotherapy concentrates on the task of employing interactive robots as therapeutic companions for people who have psychological problems or are handicapped physically, emotionally, or cognitively.

The current state of the art in robotics and in other domains within artificial intelligence is not what this book is about; it is merely the starting point for my thesis. We already have android robots,* whose appearance is designed to resemble humans, such as Honda's ASIMO, Waseda University's WABOT, and Toyota's trumpet-playing robot.† Other robots that have already been built include Volkswagen's Klaus, which can drive a car; robots that can mow our lawns and vacuum our carpets; robots that can change their own shape in order to maneuver through disaster sites in their search for victims; and robots that can reproduce—picking up and assembling the pieces of exact replicas of themselves. And already we have computer software that excels in many intellectually demanding tasks and in most areas of creativity, and we have software that can exhibit humanlike emotions.

How all these feats, and many others in AI, have been accomplished, is explained in my earlier book *Robots Unlimited,* where I also summarize the technologies that will make possible remarkable advances in the power and speed of computer processing during the decades to come, technologies such as DNA computing, quantum computing, and optical computing. When these new computer technologies have been developed to maturity, and when they have been combined with what will then be the latest advances in AI research, the intellectual capabilities and the emotional capacities of robots will be nothing short of astounding. They will look like humans (or however we want them to look). They will be more creative than the most creative of humans. They will be able to conduct conversations with us on any subject, at any desired level of intellect and knowledge, in any language, and with any desired voice—male, female, young, old, dull, sexy. The robots of the mid-twenty-first century will also possess humanlike or superhuman-like consciousness and emotions.

*Sometimes called humanoids.
†The Web site www.androidworld.com provides an extensive survey, with photographs, both of historical android projects and of current androids and domestic robots.

As the potential usefulness of robots began to be debated, alongside a discussion of the many tedious tasks that humans would delegate to machines rather than perform themselves, it was realized that the diversity of human activities needs a diversity of assistants, with different robots designed to perform and solve different tasks. Robots would be needed in industry to operate machines; they would be needed by the military and the rescue services to help at disaster sites; they would fulfill a role as replacement or adjunct teachers; they would diagnose illness and assist in the operating room. These and many other tasks soon became areas of research for roboticists.

The tasks that the early robots were designed to help solve had little to do with human emotions, and therefore they did not require any emotional response from a robot. But as psychology and cognitive science began to be studied in relation to robots, it became apparent that we need to consider what relationships might one day develop between man and machine, between human and robot. Suddenly it was important to think about what might happen when a robot communicates with a human on a personal level rather than merely for pragmatic reasons linked to the robot's "mechanical" functionality. It was no longer enough for the human to press a button or say, "Please bring me a cup of tea," and for the robot to do as requested. Instead a new generation of AI researchers was investigating more meaningful relationships between humans and what Alexander Libin has called "artificial partners."

Again it is the Japanese robot scientists who have led the research in "partner robots," recognizing that "robots increasingly have the potential to interact with people in daily life. It is believed that, based on this ability, they will play an essential role in human society in the not-so-distant future."[4]

There are those who doubt that we can reasonably ascribe feelings to robots, but if a robot *behaves* as though it has feelings, can we reasonably argue that it does not? If a robot's artificial emotions prompt it to say things such as "I love you," surely we should be willing to accept these statements at face value, provided that the robot's other behavior patterns back them up. When a robot says that it feels hot and we know that the room temperature is significantly higher than normal, we will accept that the robot feels hot. When it says that the piano is being played too loudly, recognizing of course that it is listening to a piano, we will accept that the music is too loud for the robot if it also sounds loud to us. Just as a robot will learn or be programmed to recognize certain states—hot/cold, loud/quiet, soft/hard—and to express feelings about them, feelings that we accept as true because we feel the same in the same circumstances, why, if a robot that we know to be emotionally intelligent, says, "I love you" or "I want to make love to you," should we doubt it? If we accept that a robot can think, then there is no good reason we should not also accept that it could have feelings of love and feelings of lust. Even though we know that a robot has been *designed* to express whatever feelings or statements of love we witness from it, that is surely no justification for denying that those feelings exist, no matter what the robot is made of or what we might know about how it was designed and built.

The mere concept of an artificial partner, husband, wife, friend, or lover is one that for most people at the start of the twenty-first century challenges their notion of relationships. Previously, the relationship between robot and human has always been considered in terms of master to slave, of human to machine. But with the addition of artificial intelligence to the machine-slaves conceived in the twentieth century, we have now made them into something much more. Yes, they might still be programmed to do our bidding, yet they are also being programmed to consider not only our practical wishes, serving drinks and mowing the lawn, but our feelings as well. By endowing robots with the capability of communicating with us at a level we can understand, a human level, and by building robots that have at least some appearance of humanlike features, we are rapidly moving toward an era when robots interact with us not only in a functional sense but also in a personal sense.

In the middle of the twentieth century, the founding father of the science of cybernetics,* Norbert Wiener, extolled the virtues of the

---

*Cybernetics is the science of control and communication, with an emphasis on self-controlling and self-adaptive systems—that is, autonomous systems that can learn.

interactive robots of the future, including their ability to learn from experience and, as a result of what they learn, to improve the lot of those with whom they might be interacting. The psychological benefits of such robots would, he asserted, be similar to the psychological benefits that the cared-for receive from their human carers. But to gain acceptance by the humans with whom they are interacting and for whom they are caring, robots need to imitate at least some of our social cues and to be at least vaguely similar to us in appearance.

Take a look at the Kismet robot designed and built at MIT by a team led by Cynthia Breazeal. It has a head, as we do; it has eyes, as we do; it has a mouth with moving lips, as we do. Put simply, a human interacting with a robot will be more at ease if the robot exhibits some human appearance and characteristics than if the robot is merely a metal box with wires, lights, and wheels. The more humanlike a robot is in its behavior, in its appearance, and in the manner with which it

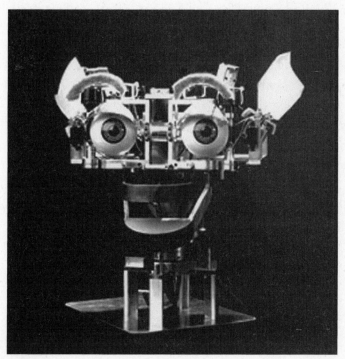

KISMET

interacts with us, the more ready we will be to accept it as an entity with which we are willing or even happy to engage.

For this reason certain trends in toy design can be viewed as precursors to twenty-first-century android robot designs. While remarkable advances were seen in robotics research during the latter decades of the twentieth century, the cosmetic appearances and forms of dolls and similar toys have been part of a less dramatic but nevertheless important trend in design. Even in that iconic product the Barbie doll, one can see breasts, while some other dolls, intended for older children and young teenagers, are marketed with a line of seductive-looking lingerie. There have also been boy-shaped characters with prominent penises, marketed as props for use in sex education.

The benefits for human-robot interaction of the human's familiarity with the robot's appearance and behavior are mirrored in the relationships between many humans and their pets. The human-pet relationship is also a kind of partnership, with some parallels to certain aspects of human-human relationships. It is a partnership that was enthusiastically seized upon by robot designers in the early days of recreational robots. In the case of traditional family pets—cats, dogs, rabbits, and the like—our relationship partnerships with those animals create a measure of emotional attachment and have been shown to be of therapeutic benefit to us. To make the partnership work with robots, designers have created robot dogs such as Sony's AIBO, robot cats, and other animal-like robots such as Furby, which sold more than 40 million pieces. Furby starts out life talking in a gibberish language called Furbish but with time reduces the incidence of Furbish in its vocabulary and correspondingly increases its use of English or whatever other language is programmed. Thus Furby enjoys a virtual kind of growth in its communicative ability. Despite this capability, Furby gave almost no appearance of being intelligent, but it was widely perceived as being cute, and not only by children. (When my wife and I gave a party during the Furby craze, some of our friends brought their children along, but it was the adults who most monopolized our Furby.)

While conducting their market research prior to designing robot pets, the most successful companies have discovered that artificial interactive pets sell better when they resemble real animals in appearance and behavior, when they simulate the experience of traditional pet ownership, thereby creating similarities that cause our perception *of* them to influence our emotional attachment *to* them. The more animal-like they are, the more attached we become. This is especially true with children, who will describe their feelings toward a pet robot in terms similar to those they employ when talking about their friends, a phenomenon known as transference.* This type of treatment of pet robots by children has been explored by Sherry Turkle's group at MIT. They described how one of their child subjects, a soft-spoken, intelligent, and well-mannered ten-year-old girl named Melanie, reacted to the robotic dog AIBO† and the electronic doll My Real Baby‡:

> Melanie believes that AIBO and My Real Baby are sentient and have emotions. She thinks that when we brought the robotic dog and doll to her school "they were probably confused about who their mommies and daddies were because they were being handled by so many different people." She thinks that AIBO probably does not know that he is at her particular school because the school is strange to him, but "almost certainly does know that he is outside of MIT and visiting another school." She sees her role with the robots as straightforward; it is maternal.
>
> One of Melanie's third-grade classmates treats My Real Baby as an object to explore and handles it very roughly, poking its eyes, pinching its skin to test its "rubber-ness" and putting her fingers roughly inside its mouth. Observing this, Melanie comes

---

*The term "transference" was originally coined in psychology to describe the process whereby a significant relationship early in one's life can be responsible for transferring one's feelings about that person to a psychoanalyst encountered later in life. For example, a patient who had a cold and distant father might view her psychoanalyst as being cold and distant. As transference theory developed within the field of psychology, so the term also came to refer to a similar phenomenon with people other than one's psychoanalyst. Recent psychoanalytic thinking has further adapted the term to apply "to relationships people have with modern technologies, especially computers."⁵ The subject of transference is discussed further in chapter 5.
†See pages 97–99.
‡See pages 16–17.

over to rescue the doll. She takes it in her arms and proceeds to play with it as though it were a baby, holding it close, whispering to it and caressing its face. When she is about to take it home, Melanie says, "I think that if I'm the first one to be with her then maybe if she goes home with another person [another study participant] she'll cry a lot . . . because she doesn't know, doesn't think that this person is its mama."[6]

Although children are, in theory, more gullible than adults, they often represent the most fundamental of human reactions. Because of this their behavior and reactions can tell us a great deal about ourselves, enabling robot designers to learn about some of the basic needs in human companionship, and as robot manufacturers have learned more about what makes a robot attractive as a companion, so different robot applications have sprung up. The most popular and hence the bestselling robots have been those produced for entertainment: Sony's robotic dog, AIBO; Honda's walking android, ASIMO, which is even able to climb stairs; robots that can tell jokes; and the 2004 bestseller, Robosapien. Possibly because electronic learning aids were so popular during the late 1980s and much of the 1990s, educational robots have also proved to be a marketing success. Other robots of which prototypes have been demonstrated include Tohoku University's ballroom-dancing androids that can predict the movements of a dancing partner, enabling these robots to follow a fellow dancer's lead without stepping on any toes. Another example is the NEC Corporation's personal robot that can recognize the faces of individual members of a family, entertain family members with its limited speech ability, and act as an interface to control the television and e-mail. A study of children aged three to five, by Thomas Draper and Wanda Clayton at Brigham Young University, found that robots with some sort of persona, robots who move and smile and say something in praise of a child's success, make better teachers and engender a greater level of motivation in their pupils than do inanimate, machinelike robots that do not talk. The My Real Baby doll, manufactured by the toy giant Hasbro, is a good example—it speaks, it makes realistic sounds, its face moves in babyish ways and it exhibits several human emotions.

The interactive aspect of a robot's being is becoming an important or even an essential element of its usefulness. Carer robots and teacher robots are just two examples. As the learning abilities of robots develop from the primitive to the sophisticated, so robots will be able to adapt to the needs and desires of their human partners. No longer will it be necessary to redesign or even reprogram a robot to perform some new task for us; instead the robots of the future will learn by watching what makes us happy and grateful and will sense our desires and satisfy them. These artificially intelligent entities will no longer be perceived as some sort of machine. Rather they will become accepted as good companions. It is the leap into this realm of relationships capable of satisfying human needs that has spawned the new disciplines of robotic psychology and robotherapy.

These new disciplines focus on the psychological aspects of our relationships with robots. While regular psychotherapists aim to help us gain some useful introspection into our own problems—both problems of self and those born out of our relationships with other human beings—robopsychology is concerned specifically with problems born of our relationships with robots. It is a highly complex minefield of new notions about relationships, in which the different ways people interact with robots and the different types of robot personality both have an effect. These effects can include caring and other therapeutic regimes, tailored by the robot designer (or, eventually, by the robot itself) to the specific needs of the individual.

Carer robots for the elderly belong to a product category that is fast attracting the interest of major manufacturers, particularly in Japan. In 2004 a "robot suit" was launched for the elderly, a motorized, battery-operated pair of trousers designed to help the aged and infirm to move around on their own. Then there is the Wakamaru, a mobile, three-foot-high talking robot equipped with two camera eyes, used mainly by the Japanese to keep an eye on their elderly parents at home. Sanyo has even developed a robot for bathing and shampooing the elderly. According to the Japan Robot Association, products of this sort will increase Japanese sales of domestic robots to $14 billion in 2010 and $40 billion in 2025 because of the accompanying marked rise in the percentage of senior citizens in the population, a rise that has created huge interest in how best to satisfy the needs of the elderly. Similar bulges in the age statistics will soon hit just about every developed country, partly because people in those countries are living longer than their parents and grandparents did and partly because of the post-WWII baby boom. As a result, robots represent one of mankind's best chances of being able to cope by providing therapeutic care for the aged.

Research into the development of robot pets for therapeutic purposes has resulted in, among other technologies, artificial fur that incorporates touch sensors, allowing an artificial pet to respond when it is stroked. This touchy-feely attribute further increases the therapeutic value of a pet when combined with a robot's lifelike appearance and behavior patterns. The pleasure of stroking a pet, together with the responses programmed into the pet for when it is stroked, have been found to enhance the experience for the elderly from both a psychological and a physiological perspective, thereby creating a friendlier mood in the patient. As a result, the moods of the patients and their overall feelings of comfort are generally improved by the stroking experience.

Several researchers and companies, particularly in Japan, have been developing the concept of robots as partners for people, and "partner robots are beginning to participate in human society by performing a variety of tasks and functions."[7] Takayuki Kanda and his team at ATR Intelligent Robotics and Communication Laboratories in Kyoto recognize the importance of finding common ground between humans and robots in order to establish relationships and to build them over time, just as normal human-human relationships evolve with time, and they have identified various goals in robotics research that will need to be achieved in order to enable robots to exhibit sufficiently humanlike behavior patterns to engender human empathy. One of these goals is for robots to recognize individuals: "It is vital that two parties recognize each other for their relationship to develop. . . . Although person identification is an essential requirement for a partner robot, current visual and auditory sensing technologies cannot reliably support it. Therefore an unfortunate consequence is that a robot may behave the same with everyone. . . . Misidentification can ruin a relationship. For example, a person may be hurt or offended if the robot were to call the person by someone else's name." Another capability that must be improved in order to facilitate smooth human-robot interaction is language communication. "Whereas speaking is not so difficult for the partner robot, listening and recognizing human utterances is one of the most difficult challenges in human-robot interaction. Although some of the computer interfaces successfully employ speech input via microphone, it is far more difficult for the robots to recognize human utterances, because the robots suffer from noise from surrounding humans (background talk) and the robot body (motor noise). . . . We cannot expect ideal language perception ability like humans. However, we believe that robots can maintain interaction with humans, if they can recognize other human behaviors, such as distance, touching actions, and visual movements, in addition to utterances."[8]

Another feature of robots deemed necessary by these Japanese researchers for successful, natural-appearing human-robot interaction is a body that looks human. "People have bodies that afford sophisticated means of expression through diverse channels. We believe that a robot partner, ideally, would have a humanlike body. A robot with a humanlike body allows people to intuitively understand its gestures, which in turn causes people to behave unconsciously as if they were communicating with a human. . . . Eye contact, gesture observation, and imitation in human-robot interactions greatly increase people's understanding of utterances. . . . Close synchronization of embodied communication also plays an important role in establishing a communicative relation between the speaker and the listeners. . . . We believe that in designing an interactive robot, its body should be based on the human body to produce the most effective communication."[9]

A recent intervention in the attempt to create humanlike robots has come from Korea, at the hands of the very same academic who invented robot soccer.* Kim Jong-Hwan has developed robot software that incorporates a computer form of DNA. Fourteen simulated chromosomes, occupying only a tiny amount of computer memory,† enable Kim's robots to exhibit up to seventy-seven human behavior patterns, which is probably more than many couch potatoes have in their repertoires. Kim's chromosomes are also intended to give robots the ability to reason and to feel desire and lust, just like us.

I fully expect that in the shorter term many of the ideas and predictions expressed in this book will be met with a certain amount of doubt, or downright disbelief, and possibly hostility. To my mind, those who doubt the possibility of computer life or robot life lack a breadth of vision similar to those who, in the 1960s, doubted the possibility of an artificial intelligence. One of the most famous outpourings of doubt expressed about AI was triggered by Berkeley philosopher Hubert Dreyfus's 1972 book, *What Computers Can't Do*. Dreyfus had previously announced, in a report for the Rand Corporation written in 1965, that artificial intelligence was a fraud, describing it as alchemy. And in 1972 he insisted, as an example of this "fraud," that "computers can't play real chess," a statement that Garry Kasparov and many other leading grandmasters now know, to their cost, to be absurd. A similar degree of skepticism has also been applied to many of the advances in scientific, sociological, and philosophical thinking through the ages. One of the best-known examples of this was Charles Darwin's theory of evolution which, in 1925, led to the famous "Monkey Trial" in Tennessee, when the renowned lawyer Clarence Darrow fought to allow Darwinism to be taught in schools. Even in the twenty-first century, there are objections being raised in some American states to such teachings.

Just as there are still those who dispute Darwinism, there will be those whose doubts and hostility toward what is written here will similarly emanate from their religious views. I do not expect the acceptance of love and sex with robots to become universal overnight. On

---

*Soccer matches between teams of robots have become a major international technical sport since its inception in 1996.
†Some two thousand bytes of data.

the contrary, it would not surprise me if a significant proportion of readers deride these ideas until my predictions have been proved correct. It is inevitable that a measure of hostility will be expressed toward such concepts, just as there was hostility toward the "ridiculous" notion that the earth is round rather than flat, toward the suggestion that the sun orbits our planet rather than vice versa, and toward the evolutionary studies that have shown man to be related to the apes. Such hostility always takes its time to dissipate, but dissipate it does. We like to think of ourselves as "special" beings—special in the sense that our consciousness raises us above every other form of life. But as psychologists, brain researchers, and other scientists learn more and more about the workings of the human mind, making them clearly explicable where now they are shrouded in mystery, then and only then will it become generally accepted that, marvelous though the human brain is, it is a kind of biological machine that can be analyzed and simulated, even to the point of simulating our emotions.

Those among you who *are* skeptical might regard some or all of my forecasts as being highly unlikely, or much further away in time than I am suggesting, or even impossible. But to take such a position would be to ignore the increasingly rapid rate of progress in artificial intelligence, materials science, and the various other relevant areas of technology. Given the dramatic technological changes and advances that the world has witnessed during the past fifty years, any assumptions of unlikelihood or impossibility regarding our technological future are at the very least risky, and most probably unjustified. Would those among you who are skeptics have believed, fifty years ago, that the accolade awarded annually by *Time* magazine for the Man or Woman of the Year would, in 1983, be given instead to the computer? And is it any more unlikely that by 2033 this same accolade will be awarded to the android—a humanlike robot?

In her groundbreaking book *The Second Self,* Sherry Turkle eloquently makes the point that we should be asking the question "not what the computer will be like in the future, but instead, what will *we* be like? What kind of people are we becoming?" *That* is where this book begins. Accepting that huge technological advances will be achieved by around 2050, my thesis is this: Robots will be hugely attractive to humans as companions because of their many talents, senses, and capabilities. They will have the capacity to fall in love with humans and to make themselves romantically attractive and sexually desirable to humans. Robots will transform human notions of love and sexuality. I am not suggesting that most people will eschew love and sex with humans in favor of relationships with robots, though some undoubtedly will. But what *does* seem to me to be entirely reasonable and extremely likely—nay, inevitable—is that many humans will expand their horizons of love and sex, learning, experimenting, and enjoying new forms of relationship that will be made possible, pleasurable, and satisfying through the development of highly sophisticated humanoid robots. This is the answer to Turkle's question "What kind of people are we becoming?" Humans will fall in love with robots, humans will marry robots, and humans will have sex with robots, all as (what will be regarded as) "normal" extensions of our feelings of love and sexual desire for other humans. Love with robots will be as normal as love with other humans, while the number of sexual acts and lovemaking positions commonly practiced between humans will be extended, as robots teach more than is in all of the world's published sex manuals combined. Love and sex with robots on a grand scale are inevitable. This book explains why.

PART ONE

> > > > > > > > > > *Love with Robots*

We ask [of the computer] not just about where we stand in nature,
but about where we stand in the world of artefact. We search for
a link between who we are and what we have made, between who
we are and what we might create, between who we are and what,
through our intimacy with our own creations, we might become.

—Sherry Turkle, *The Second Self*

1 | Falling in Love (with People)

▼ Why on earth should people fall in love with robots? A very good question, and one that is central to this book. But before we can begin to answer this question, we need to examine exactly why we humans fall in love, why love develops in one person for another human being.

Since the 1980s many aspects of love have become hot research topics in psychology, but one area that has been relatively neglected by researchers is *why* people fall in love. Even more surprising, perhaps, is the conclusion of some recent studies that romantic love is a continuation of the process of attachment, a well-known and well-studied phenomenon in children but less studied in adults. Attachment is a feeling of affection, usually for a person but sometimes for an object or even for an institution such as a school or corporation.

Children first become attached to objects very early in their lives. Babies only a few weeks old exhibit some of the signs of attachment, initially to their mothers, and as babies grow older, the signs of attachment extend to certain objects and remain evident for several years. A baby cries for its blanket and its rattle, a toddler for its teddy bear; a primary-school child yearns for her doll. Different items become the focus of each child's possessive attentiveness as the process continues, but with changing objects of attachment. Toys, Walkmen, computer consoles, bicycles, and almost any other possession can become the focus of the attachment process. As the child develops into a young adult who in turn develops into a more mature adult, so the process continues to hold sway, but with the object of focus generally changing to "adult toys" such as cars and computers. And, as the psychologists now tell us, attachment to people becomes evident in a different guise, as adults fall in love.

]]]]] Attachment and Love

Attachment is a term in psychology most commonly used to describe the emotionally close and important relationships that people have with each other. Attachment theory was founded on the need to explain the emotional bond between mother and infant.* The British developmental psychologist John Bowlby, one of the first investigators in this field, described attachment as a behavioral system operated by infants to regulate their proximity to their primary caregivers. He explained the evolution of such a system as being essential for the survival of the infant, in view of its inability to feed itself, its very limited capacities for exploring the world around it, and its powerlessness to avoid and defend itself from danger. Bowlby also believed that the significance of attachment is not restricted to children but that it extends "from the cradle to the grave," playing an important role in the emotional lives of adults.

Bowlby's notion of attachment as a phenomenon that spans the entire human life span was first explored at a symposium organized by the American Psychological Association in 1976, and during the 1970s and early 1980s Bowlby's ideas on attachment were embraced by several psychologists investigating the nature and causes of love and loneliness in adults. Some of these researchers had observed that the frequency and nature of periods of loneliness appear to be influenced by a person's history of attachment, but until the late 1980s there was no solid theory that linked a person's attachment history with his or her love life. Then, in 1987, Cindy Hazan and Philip Shaver suggested that romantic love is an attachment process akin to that between mother and child, a concept that they then applied successfully to the study of adult romantic relationships, with the spouse and various significant

*Or, more generally, between a child and its primary caregiver.

others replacing parents as the attachment figures. The principal propositions of their theory have been summarized as follows:

1. The emotional and behavioral dynamics of infant-caregiver relationships and adult romantic relationships are governed by the same biological system.
2. The kinds of individual differences observed in infant-caregiver relationships are similar to the differences observed in romantic relationships.
3. Individual differences in adult attachment behavior are reflections of the expectations and beliefs people have formed about themselves and their close relationships, on the basis of their attachment histories. These "working models" are relatively stable and, as such, may be reflections of early experiences with a caregiver.
4. Romantic love, as commonly conceived, involves the interplay of three major biological behavior systems: attachment (lovers feel a dependence on each other in a way that is similar to how a baby feels about her mother); caregiving (one lover sees the other as a child that needs to be cared for in some way); and sex (for which there is no simple parallel in attachment theory).

In practice, the similarity between infant-caregiver attachment and adult romantic attachment manifests itself principally in four different ways: Both infants and adults enjoy being in the presence of their attachment figures and seek them out to engender praise when they accomplish something or when they feel threatened; both infants and adults become distressed when separated from their attachment figures; both infants and adults regard their attachment figures as providing security for them when they feel distressed; and both infants and adults feel more comfortable when exploring new possibilities if they are doing so in the presence of, or when accessible to, their attachment figures.

Hazan and Shaver's theory of romantic love as an attachment process contributed little to psychologists' understanding of the *role* played by attachment in romantic relationships, or to how that form of attachment evolves. Shaver's view at the time was that the process of natural selection had somehow "co-opted" the human attachment system in order to facilitate the bonding process in couples, thereby promoting feelings akin to the parental instincts that help infants to survive. But during the 1990s, researchers into the theory instead began to come to the conclusion that there exists a "modest to moderate degree of continuity in attachment style"[1] as a person ages, implying that those infants who have strong attachment bonds with their mothers are more likely to grow into adults who have strong attachment bonds with their partners. If this is indeed the case, then one's *capacity* to experience romantic love would appear to depend on one's attachment history.

Attachment to a material possession can develop into a stronger relationship as a result of the possession's repeated use and the owner's interaction with it. This phenomenon is known as "material possession attachment."[2] The process by which this happens is similar to the way in which we develop our understanding of and feelings for people as we get to know them over time. Initially, of course, a material possession is nothing more than a commodity that is purchased and probably comes to "live" in our home. As we use it, play with it, and so forth, we get to know it, and gradually it might become less and less of a commodity, more and more a part of our life. The computer is no longer simply *a* computer, it quickly becomes *my* computer. Not so much "my" in the sense of its being owned by me, but more in the sense of its being the particular computer with which I associate myself, the one that I feel is part of my being. Computers, in fact, provide an excellent example of this interpretation of "my"— when people go into an Internet café or into the computer room at school or college, they will usually gravitate toward the same computer they have used in the past, even though all the machines in the room might be, to all practical purposes, identical. They head straight for "their" computer, the one for which they feel they have some affinity, the one with which they subconsciously feel they have already developed some sort of relationship.

As an owner uses an object and interacts with it more and more over time, so this personal attention applied to the object endows it with a special meaning for the owner. Several psychology researchers have pointed to this creation-of-meaning process, among whom Mihaly Csikszentmihalyi and Eugene Rochberg-Halton have been the primary advocates, referring to this special meaning as "psychic energy." As the owner invests more psychic energy in an object, more meaning is attached to the object, it becomes more important to its owner, and the stronger is the attachment that the owner feels for the object.

The commodity thus becomes increasingly personalized to its owner through repeated use and interaction, and as it does so, it takes on, within the owner's mind, an aura of uniqueness. Consciously the owner knows full well that his computer is more or less exactly the same as millions of other computers in the world, but subconsciously there develops in the mind of the owner the notion that this particular computer, *his* computer, is unique, it is personal to *him*. And now that the commodity is no longer viewed as a commodity but as something unique, something personalized, it becomes part of its owner's being, "symbolizing autobiographical meanings."[3] The computer, if that is the commodity, becomes irreplaceable in the mind of its owner, even though clearly it could be replaced by another computer of the same make and model with the same amount of memory and the same operating system.* This "uniqueness" will often cause the owner to be unwilling to replace it, "even with an exact replica, because the con-

---

*Throughout this book, when discussing the interaction between a user and a computer, I employ the word "computer" to mean the combination of the computer hardware (the box, the keyboard, the mouse, and the screen) with whatever software it is running (the programs that make the computer do clever things). What the user actually interacts with is the software. The computer keyboard, the mouse, the text on the computer screen, and any speech output that the user hears are all merely the means by which the user interacts with the software. The software itself is invisible, leading the user to talk about his interaction as being with the computer rather than with the computer-software combination.

sumer feels that the replica cannot sustain the same meaning as the original."[4] Such possessions thereby become endowed with personal meaning that connects the object with its owner—the object in a sense becomes part of the owner—and this personal meaning is what is called "material possession attachment."

There are of course many reasons an owner could develop a sentimental attachment to a particular object, but these reasons normally derive from something connected with the source of the object— perhaps it was a gift from a loved one, a memento of an emotionally important event in the owner's life, or a personal possession the owner has used caringly for several years. What is different about the nature of the possession attachment felt for a computer is the element of control—the computer is at its owner's every beck and call. Russel Belk's 1988 paper "Possessions and the Extended Self" discusses the notion that we are "extended" by our possessions, they become part of us, extending us, whether they be material possessions or human "possessions" such as "my" friend, "my" partner, "my" spouse; and Beck cites David McClelland's suggestion that the greater the control we exercise over an object, the more closely allied with that object we become.

Thus, through the great measure of control we exercise over computers, we have the potential to become close to them. Because of the high level of use we make of them and the interactive nature of that use, computers have the potential to hold a special meaning for us, to strengthen the attachment we feel for them. Combine these with the potential to extend ourselves by virtue of our possessions and it is not difficult to imagine that the computer—controlled, interactive, used, and possessed—could create in us the level of attachment necessary to engender a kind of love. And if, as suggested by Frayley's thinking, one's *capacity* to experience romantic love depends on one's attachment history, an attachment history that involved computers or electronic pets could provide a basis for the *capacity* to fall in love with robots.*

*We shall return to the subject of attachment in chapter 3.

FALLING IN LOVE (WITH PEOPLE)

]]]]] How Proximity and Repeated Exposure Affect Falling in Love

There have been a number of studies on the effect of proximity on attraction. In one of the earliest studies, conducted during the 1930s in Philadelphia, the addresses of marriage partners were recorded for some five thousand marriage licenses. It was found that 12 percent of the couples lived in the same building at the time they applied for a marriage license* while a further 33 percent lived within five blocks of each other. For a similar study, this one in Columbus, Ohio, during the 1950s, the investigators interviewed 431 couples and found that 54 percent of them lived sixteen blocks or less apart when they first dated, and for 37 percent of these couples the distance was five blocks or less. Surveys at MIT and the University of Michigan found similar results for couples living in student dormitories. The MIT study showed that the most important factor in creating emotionally close couples was the distance between their apartments—the closer they lived, the more likely they were to become friends, while the University of Michigan study indicated that roommates were much more likely to become close friends than were students living in different rooms several doors away from each other.

The overwhelming conclusion to be drawn from these and many similar studies is that seeing someone frequently, referred to by psychologists as "repeated exposure," creates a much more fertile atmosphere for friendship and love than seeing someone less often, and the proximity of their living quarters clearly has a significant effect on how frequently two people meet. If two people live close to each other, they are more likely to develop a familiarity than if they live farther apart— familiarity in terms of seeing each other more, spending time with each other, thinking about each other, and anticipating interaction with each other.

It has also been shown that even without any personal contact

---

*Given the social mores of the time, the vast majority of these couples would not have been living together but would instead have been living in different apartments in the same building.

with the other individual, repeated exposure *to* them generally creates a feeling of liking *for* them. The reason that repeated exposure appears to create such a positive effect on human attraction has been suggested by Ayala Pines to "arise out of an inborn discomfort that we all feel around strange and unfamiliar things."[5]

In an experiment conducted by Richard Moreland and Scott Beach at the University of Pittsburgh, four women pretended to be students attending classes. The women avoided all contact with the other students in the class, and they attended different numbers of lectures: One of them attended once, another ten times, one fifteen times, and the fourth one not at all. At the end of the course, the students in the class were shown photographs of all four women and asked about their feelings and attitudes to each of them. Even though none of the students had had any personal contact with any of the four women, their reported liking of each of the women was strongly related to how often that woman had attended the class—the one who never attended was liked the least, with the level of liking rising as the number of attendances in the class rose. The study also found that the more often a woman attended the class, the more likely she was to have been described by the students as attractive, interesting, intelligent, and similar to themselves.

The common factor in the studies described above is that in each case the repeated exposure was to another person, but Robert Zajonc has shown that repeated exposure to almost anything increases our tendency to like it and that a direct correlation exists between the frequency of exposure and the level of liking. In one of his experiments, he pretended to be conducting a test of visual memory and asked his test subjects to look at photographs of different people, with each viewing lasting for thirty-five seconds. He varied the number of times each photograph was shown—some were shown once while others were shown two, five, ten, or as many as twenty-five times. Zajonc found that his subjects tended to feel more positively toward the person in a photograph if they were shown their photograph more often, indicating that the physical *presence* of the object of one's affection is not a prerequisite for developing that affection. This result concurs with the phenomenon of pen pals falling in love without meeting, and also with its more recent and more prolific parallel—falling in love on the Internet.*

]]]]] Why People Fall in Love

Let us now retrace our steps a little, from falling in love as a form of attachment to a discussion of this question: Why does Human X fall in love with Human Y rather than with Human Z? What is it about our partners that causes us to fall in love with *them*?

We like or dislike a person according to how we feel in that person's presence—"like" is a feeling for someone in whose presence we feel good. The extent to which we are attracted to someone has been found to depend on the number of positive and negative feelings we have toward that person and was expressed by Donn Byrne in 1971 as "Byrne's Law of Attraction," derived from investigations into the attraction feelings of students who volunteered to take part in psychology experiments. The formal expression of Byrne's law does not exactly make for romantic reading—it looks like this:

$$\text{Attraction} = m \times \frac{\Sigma \, positive \, feelings}{(\Sigma \text{ positive feelings} + \Sigma \text{ negative feelings})} + k$$

What this means is that the strength of attachment one feels for another person is governed by the strength of one's *positive* feelings for that person in relation to the strength of *all* of one's feelings for them. Here are two very simple examples: Let us assume that you have feelings about ten different facets of a particular person—their character, their looks, their personality, their conversational style, and so on—and that the feelings you have about each of these ten facets are identical in strength. If you have good feelings about eight of those ten facets

*See the section "Falling in Love on the Internet," later in this chapter.

and bad feelings about two of them, then the fundamental measure of attraction that you experience toward them is

$$8 / 10$$

because you have eight positive feelings for them out of a total of ten feelings (positive and negative). But if you have good feelings about only three of the person's facets and bad feelings about the other seven, then the fundamental measure of attraction that you experience will be only

$$3 / 10$$

This somewhat cold mathematical approach to the magic of human attraction might appear, to the more romantically inclined reader, to be utter nonsense and just about as far removed as one could imagine from reality. But in fact the accuracy and usefulness of Byrne's law has been proved in many psychology experiments since he first stated it—experiments in which various positive and negative emotions were manipulated and where the calculation of the "attraction" measure produced the values that the experimenters predicted. And although Byrne's results stem from the experimental-psychology laboratory and are derived from first impressions of one person about another, Byrne discovered "that the same factors found to operate in the laboratory are also found to operate in determining real-life friendship, love, courtship and marriage."[6]

Interestingly, Byrne's Law also shows that we are more inclined to like someone when we are experiencing positive feelings for reasons that might not be associated with that particular person but which are causing the same feelings in *them,* such as both hearing good news ("We've passed!") or listening to music that they *both* enjoy. Conversely, it has been discovered that two people will tend to be less attracted to each other, or even to dislike each other, if they are sharing negative feelings, such as "We've failed" or listening to music that both hate.

A simpler way of inducing people to fall in love was investigated by a team led by Arthur Aron at the University of California at Santa Cruz. In 1991, Aron experimented by taking pairs of students who had never met, putting them in a room together for ninety minutes, and asking them to exchange intimate information, such as their most embarrassing moment and how they would feel if they lost a parent. Immediately following this part of the experiment, they were asked to stare into each other's eyes for two minutes without speaking. At the end of the experiment, the two subjects left the room through different doors, in order to remove any possible feelings of obligation to see each other in the future. (Despite this cautionary ploy, the very first couple that took part in Aron's experiment were married six months later.) All the students in the experiment were asked to rate the closeness of the relationship formed within their pair at different stages of the ninety-minute period, and the ratings were compared with those of a group of similar students who were asked to rate the closest relationships in their lives. A key result from the experiment was that after only forty-five minutes of interaction the relationship between the paired students was rated as closer than the closest relationship in the lives of 30 percent of similar students. Although there might have been some bias among the paired students when giving their "closeness" ratings, due to the fact that they knew they were involved in an experiment, this 30-percent figure suggests that self-disclosure can be a powerful and fast-acting device in getting someone to feel attracted to you.

Talking intimately about one's most embarrassing moments and baring one's emotional soul as means of engendering affection from another person could prove to be a double-edged strategy. If a robot tried this on someone who was not in the mood to reciprocate, the response from the human might be to suggest that the robot needed therapy or that its software or hardware needed fixing. But the strategy works well when the behavior *is* reciprocated, because the other person will understand the emotional risk that lies in emotional self-disclosure, and if he or she is willing to share that risk, then the mutuality of the risk will likely become a bonding agent. It is well established that couples who experience the risk of physical danger together—for example, being in the same vehicle in a traffic accident—tend to bond strongly and swiftly.

]]]]] Measuring Love

Neurobiologists Andreas Bartels and Semir Zeki of University College London reported in 2000 on an analysis of fMRI scans* of the brain activity of love-struck students† while they were gazing at photographs of their loved ones. In these cases, the brain-activity pattern was very different from when the same students looked at photographs of close friends with whom they were *not* in love. Bartels and Zeki also compared these scans with those taken of people in different emotional states and found that the pattern corresponding to romantic love was unique.

Helen Fisher, an anthropologist at Rutgers University, carried out a similar experiment in collaboration with Arthur Aron, in an attempt to find results that supported the work of Bartels and Zeki. Her team analyzed the brain scans taken of seventeen recently smitten college students, ten women and seven men, whose ages ranged from eighteen to twenty-six and who had been in love, on average, for some seven months and whose feelings of love were at a more intense level than in the participants in the Bartels-Zeki experiment. The scans for each student were taken over a forty-five-minute period, during which the subjects were shown photographs of their loved one alternating with those of a familiar acquaintance of the same age and sex as their beloved but in whom they had no romantic interest. The scans showed that the experience of romantic attraction activated those pockets of the brain with a high concentration of receptors for dopamine, a chemical closely associated with states of euphoria, craving, and addiction.

The uniqueness of the "in-love" brain scans could serve as the basis for robots to determine whether or not a particular human was falling in love with them. A robot who wants to engender feelings of love from its human might try all sorts of different strategies in an

---

*Researchers know the general area of the brain where various functions occur, such as speech, sensation, and memory. An fMRI (functional magnetic resonance imaging) scan provides a picture of the brain and helps to determine *precisely* which part of the brain is dealing with certain functions.
†The average length of time these students had been in love was twenty-nine months.

attempt to achieve this goal, such as suggesting a visit to the ballet, cooking the human's favorite food, or making flattering comments about the human's new haircut, then measuring the effect of each strategy by conducting an fMRI scan of the human's brain. When the scan shows a higher measure of love from the human, the robot would know that it had hit upon a successful strategy. When the scan corresponds to a low level of love, the robot would change strategies.

]]]]] Ten Causes of Falling in Love

The first systematic study of why someone falls in love with a particular person was published in the *Journal of Social and Personal Relationships* in 1989. Arthur Aron, Donald Dutton, Elaine Aron, and Adrienne Iverson modeled their study mainly on three earlier accounts of falling in love, obtained from surveys conducted by other psychology researchers.* One of these surveys was based on detailed written accounts of falling in love by students who had done so during the preceding eight months. A second study compared the experiences of two hundred attendees of a seminar titled "Love and Consciousness," who wrote accounts of their experiences of falling in love or falling "in friendship." For the third study, a questionnaire was compiled to investigate the subjects' most recent experiences of falling in love and, in particular, the moment when they first experienced a strong feeling of attraction.

A review of the reports on these surveys reveals eleven factors that appear to be major contributors to the process of falling in love. One of these factors, proximity, is an explanation of why people come to be in a *situation* that engenders love rather than a factor that causes love to develop when they *are* in that situation, and for this reason I have not included a discussion of proximity in this section.† We therefore have ten factors to consider, and in chapter 4 we shall see that

---

*Lawrence Belove (in 1980), Philip Shaver et al. (in 1987), and Dorothy Tennov (in 1979).
†However, proximity can lead to being alone with the love object, which *is* one of the ten causes.

most of these factors are equally applicable for engendering love, by humans, for robots.

1. Similarity

There is strong empirical evidence that people tend to like other people who are similar to themselves in one or more important aspect. It might be a similar level of education, similar attitudes, a common interest, a similar family or religious background, similar personality traits, similar social habits, or similarity in any of a host of other characteristics.* Similarity is thus one of the dominant reasons for initial feelings of romantic attraction. The first study to examine this phenomenon was carried out by Sir Francis Galton during the 1880s. His findings, and those of later psychologists, concluded that couples tend to be similar in all sorts of different traits: psychological traits, physical traits, and personality.

There is evidence from psychology research that we like people better when they change to *become* similar to us, as compared to when they are consistently like us. This can happen because those who change *in order* to make other people happy are often perceived as being "nicer" than those who *always* try to make other people happy— it is the gaining that promotes attraction here, the earning of esteem rather than experiencing it from the first encounter.

2. Desirable Characteristics of the Other

Most of the studies of romantic attraction have revealed, unsurprisingly, that personality and appearance are two of the most important factors in engendering a feeling of attraction. Ayala Pines found that more than 90 percent of the men and women she interviewed about the factors that caused them to fall in love mentioned a characteristic of their partner's personality, with women mentioning personality traits as a crucial factor slightly more often than men. But when it came to appearance, 81 percent of men said that they were attracted to the

*How well people's similarities match, how well they "fit together," is not only important in bringing them together, it is also a key factor in how gratifying their relationship will be and how long it is likely to be sustained.

physical appearance of their loved one, while only 44 percent of the women interviewed said that they were attracted by the appearance of their man. Given the importance of appearance in the attraction process, it is easy to understand why sex-doll manufacturers choose sexually alluring appearances for their dolls, as we shall see in chapter 7, a policy that sexbot designers will inevitably follow.

3. Reciprocal Liking

Knowing that one is liked by the other appears to be one of the dominant factors in falling in love. This factor is emphasized in Shaver's adult-attachment theory, in which the loved one (read "the cared-for" one) perceives themself to be loved by the love giver (read "the primary carer"), as a result of which the loved one knows that they are likable, which makes them feel good. And when we feel good in the presence of a particular person we are more likely to develop feelings of attraction toward them. One test of this factor came from Arthur Aron's experiment described earlier,* which had a secret ingredient added. He told both people within a couple that the other one would like them. "That expectation had a huge effect," said Aron. "If you ask people about their experience of falling in love, over 90 percent will say that a major factor was discovering that the other person liked them."

4. Social Influences

General social norms usually have a significant effect on falling in love, by screening out at an early stage some possible candidates for affection. A simple example is age—it is a social exception rather than the norm for someone to fall in love with a person who is very much older than oneself, so even if someone finds a much older person interesting or attractive, the thought will already be in their mind, "What would people think about me if I pursue a relationship with this person or accept their advances?" Similarly, some cultures screen out many candidates for affection on racial grounds, with the result that a couple who might otherwise be candidates for falling in love will often eschew

*See the section "Why People Fall in Love," page 33.

any form of relationship because one or both of them knows that it would be unacceptable in their culture. As Ayala Pines explains, "social norms reward people who follow the norm and punish those who deviate, as, for example, when friends and relatives shun or express outright criticism of an unsuitable, potential partner."[7] Pines's comment points to another way in which social norms are often influential—the social approval or disapproval of those in one's own social network, especially one's friends, can be an influence on whether or not one falls in love with a particular person, even if these influences are not culturally or racially biased.

5. Filling Needs

One of the stronger reasons for falling in love is need—the need for intimacy, for closeness, for sexual gratification, for a family. In some cases the need can be for recognition from others—a gain in status, garnered as a result of having acquired a trophy partner. So when someone says "I love you," what they might actually mean is "I need you," their subconscious hiding from them the true reason for the feeling they have developed for the object of their "love."

6. Arousal/Unusualness

The situation in which one meets a potential love object can have a significant effect on whether a feeling of attraction develops. If one is aroused, even in a negative way, by the situation itself, that arousal can have a positive effect on one's feelings of attraction. Danger is one well-known example of this phenomenon.*

7. Specific Cues

The object of one's love might possess some particular characteristic that creates an unusually strong feeling of initial attraction, such as a voice that one finds very appealing, or a physical feature, like the face, the eyes, or the shape of the body. These cases often give rise to "love at first sight."

*See the final sentence of the section "Why People Fall in Love," page 35.

8. Readiness for Entering a Relationship

Some emotional states make us much more susceptible to falling in love than do others. If we are suffering from particularly low self-esteem because our partner has just dumped us, we are ripe for starting a relationship "on the rebound." And a temporarily lowered level of self-esteem for other reasons can similarly be assuaged by a new relationship. Here again there is a need, but this time it is a need for the relationship itself rather than for what it might bring us.

9. Being Alone with the Love Object, or Exclusiveness

This is a stronger form of the factors described in the earlier section "How Proximity and Repeated Exposure Affect Falling in Love." Being alone with the object of one's love is likely to enhance those feelings of love and encourage any feelings of reciprocity that might exist in one's love object.

10. Mystery

A person who carries an air of mystery or intrigue will be often be found to be romantically appealing. Similarly, a mysterious situation can have a catalytic effect on a relationship in much the same way as danger does.

While these ten reasons are still fresh in your mind, just pause for a moment and ask yourself this question: Which of these reasons, if any, would not apply if the object of one's potential love were not another human being but instead a robot? You might ponder this question until we discuss it further in chapter 4.

]]]]] Falling in Love on the Internet

Falling in love via the Internet has become a widespread social phenomenon. This is the modern-day, vastly speeded-up version of falling in love with a pen pal you've never met, which sometimes used to happen in the days when the postal system rather than the Internet was the most popular method of written communication between those in faraway locations. An interesting aspect of Internet relationships, as with pen-pal relationships, is that some of what are normally regarded as being the most important factors in the initial attraction of one person to another—such as looks, age, and voice—are entirely missing from the initial stage of most Internet relationships.* Those involved in Internet chat and Internet flirting are usually hidden from view, hidden from hearing, and able to give the impression of being any age they wish, with the result that relationships sometimes develop between couples who, if they saw each other in a restaurant or across a dance floor or a room at a party, might never have shared a second glance. As a cartoon in *New Yorker* magazine explained, "On the Internet, nobody knows you're a dog."

This invisibility brings an important extra element to the flirting process on the Internet, as explained by Deb Levine, author of *The Joy*

"On the Internet, nobody knows you're a dog."

*This is not the case if both parties decide from the outset to use webcams and speech-transmission technology, but at the present time these are employed in a small minority of early Internet relationships.

of Cybersex: A Guide for Creative Lovers and the developer and therapist of the Columbia University Web site Go Ask Alice, where she dispenses advice on safe sex and healthy relationships.

> The online world gives those people who do not fit a stereotypical model of human beauty a chance to be Don Juans and Carmen Mirandas and have an equal opportunity to be found desirable. For those considered beautiful by societal standards, it gives them a chance to be attractive to others for reasons other than their physical qualities (i.e., intellect, charm, interests, etc.).[8]

Being attractive to others is, of course, one of the keys to a successful relationship, and it will be important for a human involved in a developing relationship with a robot to be shown and to believe that the robot is attracted to them. The fact that attraction for reasons of intellect, charm, and the like occurs so often in Internet relationships is a strong indication that humans *will* be convinced by their robot's indications of attraction and love for them. There is little point in programming a robot to tell obviously plain or ugly people that it finds them physically attractive, as the robot will lose credibility from any human partner who has the wit to detect the lie. But there is considerable point in programming a robot to search out, comment favorably on, and interact with those characteristics of the human partner that could reasonably be described as positive attributes. If the robot is programmed (or learns) to virtually enjoy the same tastes in literature, music, sports, and so on as its human partner and to appreciate its partner's personality, then it will be convincing in its appreciation for its human, and this appreciation will act as a catalyst in developing the relationship further.

Levine draws other parallels between attraction in Internet relationships and attraction in face-to-face relationships, parallels that can extend also to human-robot relationships. We have seen that proximity is an important factor in promoting attraction. Levine points out that

> In the online world, proximity is not defined by physical location, but instead by a particular chat room, message board (Internet

forum), listserv* or type of Internet software that users have in common. In order for people to meet online, they have to be in the same chat room at the same time (closest approximation to "real life" proximity), post messages on the same message board. . . .

And she recommends, for those seeking someone on the Internet who might find them attractive, that

your best bet is to find a community that revolves around a sub-ject in which you are interested (for instance, sports, health, chil-dren, books, movies) and spend time there on a regular basis.

Thus Levine demonstrates that the Internet version of proximity is also a means to take advantage of another of the principal causes of falling in love—similarity. The very nature of the Internet facilitates the process of finding similarity, allowing someone who is passionate about origami or cross-country skiing to discover a host of like-minded candidates for their affections, candidates for whom they themselves might be considered emotionally attractive. The technology of the Internet focuses the atten-tions of other origami aficionados on the user who is seeking affection, providing the opportunity for any of them who wish to do so to flirt.

The human-robot relationship takes this process of finding simi-larity an important step further. Not only will the robot be programmed and learn to have similar interests and other characteristics as its human owner, it can also be guaranteed by its programming to find its owner emotionally attractive. Instead of a user visiting a Web site where there will almost certainly be many like-minded people, but with the risk that none of them might find the user attractive, the user's robot will be both like-minded *and* attracted to the user.

Those who develop strong emotional ties on the Internet, leading to romantic relationships, constitute only a relatively small percentage of the online population. But because of the total size of the online

---

*Listserv is a leading e-mail list-management program that facilitates the administration of various types of e-mail lists, such as discussion groups.

population, even a modest minority can represent several million people. Nicola Döring quotes a telephone survey conducted in the United States in 1995, in which 14 percent of those questioned and who had access to the Internet reported having become acquainted with people on the net whom they would refer to as "friends," though no distinction was drawn between romantic and nonromantic relationships. Döring also refers to surveys aimed at people who were active in newsgroups—within this category the portion of those who maintained close relationships on the Internet was 61 percent (of which 53 percent were friendships and 8 percent were romantic relationships). The reason for the significantly higher percentage among the newsgroup members is that because a newsgroup is highly focused to a specific interest, its members are by definition similar, in that they share an interest in the group's topic. Thus a similarity of interests is a powerful factor in the generation of romantic attachment via the virtual world of the Internet, just as it is in the physical world.

The data referred to by Döring is already several years old, and since then the statistics have shot up. Cyberromance is an experience that has grown phenomenally within the Internet population, an experience whose popularity is still growing rapidly. Esther Gwinnell, in her 2004 book *Online Seductions,* points out that online relationships, not only those formed on matchmaking sites but also those that start in chat rooms and through instant messaging, have become so common that many psychotherapists in the United States now devote their practices solely to dealing with the problems caused by cyberromances. These problems include detrimental effects on preexisting relationships, especially in marriages where a spouse will often refuse to admit that a cyberromance constitutes a form of cheating.

On the positive side, Deb Levine points out that

for some people, online attraction and relationships will become a valid substitute for more traditional relationships. Those who are housebound or rurally isolated and those who are ostracized from society for any number of different reasons may turn to online relationships as their sole source of companionship.

> > > > > > > > > > >

2 Loving Our Pets

]]]]] The Nature of Human-Pet Relationships

Pet ownership is known to date back to Paleolithic times. A twelve-thousand-year-old tomb, found in Ein Mallaha in northern Israel, contained the remains of an elderly woman buried together with those of a puppy dog. The woman's left hand was placed so that it rested on the dog's shoulder, providing visual evidence for a special relationship between early humans and the animal world, which is very rare in an archaeological site. Simon Davis, a member of the discovery team, explains, "This case at Mallaha is quite clearly a rather special and almost unique example of an animal skeleton buried with a human. So I think this really points to almost a kind of emotional or affectionate relationship between the old woman and the puppy."[1]

Relationships between pets and humans have evolved considerably from the times when the roles of pets were principally as workers. Cats were originally brought into homes in many countries because of their penchant for catching mice; dogs have long been employed as hunting partners and house guards; while horses, in addition to being the fastest mode of transport for thousands of years, have also been given a variety of jobs that involve pulling heavy objects such as coaches and plowshares. But just as robots have evolved from assembly-line machines to companions for the elderly, so pets have also evolved into our companions.

Many people own pets, and a significant proportion of pet owners love their pets, spending considerable amounts of money on their food, health care, and sometimes their grooming.* In the United States, it is not at all uncommon for pets to be dressed in designer fashions, enrolled in day-care centers, given kidney transplants (and other high-tech operations) at a cost of approximately $6,500 per kidney, and to be laid to rest in pet cemeteries. In addition, pet owners put up with all sorts of inconveniences contrived by their pets, as they scratch the furniture, claw the carpets and bedding, and leave smelly deposits on the floors. Given these disadvantages of pet ownership, it seems clear that the level of attachment between pet owners and their animals is extremely high. Edward Rynearson explains this on the basis that "the human and pet are significant attachment figures for one another. Under normal circumstances they share complimentary attachment because of mutual need and response."[2]

In a paper aptly entitled "Why Do People Love Their Pets?" John Archer, a psychologist at the University of Central Lancashire, discusses the reasons people keep pets, concentrating on the most popular animals—cats and dogs. Archer's findings include the observation that in Western societies the relationship between humans and pets has intensified since the Second World War. A survey conducted by the American Pet Products Association in 2004 bears this out, indicating that 63 percent of U.S. households had at least one pet, comprising 77 million cats; 65 million dogs; 17 million birds; 16 million "pocket pets" such as rabbits, ferrets, and rodents; and even 9 million reptiles, figures that are steadily increasing by some 3 to 5 percent annually.

Several research psychologists have carried out systematic studies of love for pets, mostly positing this love in terms of attachment. Aaron Katcher led a 1983 study that investigated various common indicators of affection for pet dogs, such as talking to the dog frequently. A survey conducted among the clients of a veterinary clinic found that 67 per-

*I should perhaps confess to some bias on this subject, having lived with as many as four cats at the same time, all of whom slept on our bed, ate mounds of fish and chicken, and were whisked off to the vet at the slightest indication of illness. Sadly, Ginger, Muffin, and Smoky have all died fairly recently, aged between eighteen and nineteen (in human terms, ninety to ninety-five), and were duly cremated, their ashes lovingly scattered in the garden. Fred is still alive and well at the time of this writing.

cent had a photograph of the dog, 73 percent allowed it to sleep in their bedroom, and 80 percent believed that the pet was sensitive to the owner's feelings. In a study by *Psychology Today* magazine of more than 13,000 pet owners, 25 percent celebrated their pet's birthday. And a study by Victoria Voith found that 97 percent of 1,500 pet owners in a survey confessed to talking to their cat or dog at least once a day, while 99 percent of the owners considered their pet to be a member of the family. In other studies the "member of the family" figures have varied, from 68 percent up to 93 percent.

Most children talk to their pets and feel that their pets reciprocate their own love. Many adults, too, form strong emotional attachments to their pets, some insisting that their animal is "almost human" (despite ample evidence to the contrary provided by the pet's nonhuman appearance), and some deriving even more satisfaction from their pet relationships than they do from their social relationships with people.

]]]]] The Anthropomorphism of Pets

Much of the research into human-pet relationships has been based on anecdotal evidence and on observations by psychologists and vets. But it is also interesting and important to consider how pet owners themselves perceive and evaluate their relationships with their animals. By gaining an understanding of the owners' perceptions of such relationships, we can better assess how human relationships with robots are likely to develop.

The pioneering research into relationships between pets and their owners was led by Julia Berryman in the mid-1980s. Berryman's team found that while there was a wide variation between the pet owners in the study as to the importance they attached to their relationships with their pets, one common factor was dominant: Pet owners perceive their relationships with their pets as being more similar by far to their relationships with children, particularly in those cases where the child was their own, than they were to their relationships with their spouse or partner or with a friend. The reasons appear to be that children and pets bring similar emotional rewards, and both chil-

dren and pets depend on adults—their "owners"—especially for playing games and having fun. And just as playing games and having fun
are shared activities that bind both human-human and human-pet
relationships, other shared activities—even boring, routine activities—
tend to bind both types of relationship.

The human tendency to project feelings and thoughts onto animals would seem to be a pervasive one. It is probably based on what
developmental psychologists call "the theory of mind," the ability to
impute a mental state to others. Most humans attribute others with
having minds—that is, feelings, beliefs, and intentions different from
their own. But in making such attributions, these humans tend to over-
attribute, and in the case of animals this leads to anthropomorphism.
From the Greek words *anthropos,* meaning "man," and *morphe,* meaning "form" or "structure," anthropomorphism is a tendency to regard
and describe objects, animals, and even natural phenomena such as
the wind and the sea in human terms, attributing human characteristics to them with the intention of rationalizing their actions. Anything
that bears some similarities to a human being, and with which a person
has repeated interactions, is treated as if it has a mind. Thus an animal,
alive, affectionate, and warm-blooded, comes to be treated in certain
ways as though it were human (alive, affectionate, and warm-blooded),
leading many people to interact with their pets as if they were humans
and to form relationships with their pets that come to be like those
formed with humans.

Pet owners extend this anthropomorphism toward their animals
in other ways, including giving them individual names, feeding them
from their own plates at mealtimes, taking them to a medical practitioner when they're ill, celebrating their birthdays, allowing them to
sleep on the owners' beds, and even on occasions dressing them up like
humans. By such actions the owners cement the perception of a
humanlike relationship with their pets, but clearly, since pets are
unable to carry on a conversation with their owners, the form of love
felt by a pet owner for their animal is much closer to the form of love
that humans feel for babies than it is to a feeling of romantic love.

In two studies based on students' perceptions of the cognitive abilities of animals, Jeffrey Rasmussen and Donald Rajecki found that although the students appreciated that the cognitive abilities of dogs and humans are at different levels, they believed that the mental processes giving rise to these abilities are broadly similar, that dogs think like we do, just not so well. Pets normally live in the home and are therefore in regular or even constant proximity to their owners and to members of the same household. For this reason, pets are themselves members of the household, even if they are not always treated as part of the family, and in most households pets are perceived as individuals with their own life histories, their own personalities, and their own "personal" tastes. This individuality is a major factor in explaining why most pets *are* regarded as members of the family.

Not everyone understands the appeal of pets and the strength of the bond that is often developed by a pet owner for an animal. Some of my friends and acquaintances kid me about my devotion to my cats, in a few cases going so far as to suggest that I'm crazy. But such a love is not a phenomenon that deserves to be pilloried. As James Serpell has argued, attachment to a pet is too widespread a phenomenon throughout history and in the modern world for it to be viewed as an abnormal response by inadequate individuals.

Research into the anthropomorphism of animals has revealed that not only are pet owners more likely than nonowners to attribute humanlike understanding to their own pets, they are also more likely than nonowners to make the same attribution to animals in general. For one study, Margaret Fidler, Paul Light, and Alan Costall showed students a series of videotaped sequences of dogs in everyday settings and then questioned the students about the dogs' behavior. The common factor in the taped sequences was that the dog and its owner were interacting in some way: The owner was stroking the dog, eating with the dog at her feet, teasing the dog while talking to it, and leaving the room while the dog was watching her. The students' descriptions of the events shown in the videotapes were then classified in one of three ways: as anthropomorphic (for example, "The dog watched the person eating and moved to a position to get eye contact and sat and tried without break to get the person's attention"), or using "as if" terminology (e.g., "The dog appears to get excited. . . . He turns around as if he is looking for what the owner is talking about"), or mechanistic—descriptions devoid of any mention of meaning or purpose on the part of the dog. Those students who were or had been pet owners were found to be significantly more likely than nonowners to respond that the dogs' actions were deliberate and that their behavior resulted from their *understanding* of the situations portrayed in the video.

Some pet owners subconsciously take the process of anthropomorphism even further and describe feelings toward animals that indicate they value these relationships more than human ones. In modern Western societies, human relationships often produce difficulties and dissatisfaction, providing one reason that this may be so. Surveys of veterinary practitioners in the United States indicate that some pet owners would rather lose their spouse than their pet. Further evidence of the preference for a pet relationship over a human one comes from a 1990 survey for which Peter Peretti interviewed 128 senior citizens in a Chicago park and found that they devoted considerably more time to describing dogs as friends than to describing people as friends. In fact, 75 percent of the men and 67 percent of the women in Peretti's survey said that their dogs were their *only* friends. And in a study by Sandra Barker and Randolph Barker, it was found that on average the owners felt significantly closer to their dogs than they did to other (human) members of their family.

More recent research supports these findings. In one sample quoted by Archer, taken from a questionnaire study of dog owners, more than half of those surveyed agreed that the loss of their dog would mean as much to them as the loss of a family member or friend. Some owners also made favorable comparisons with human relationships, typical of which were these: "I care for them more than for most people I know," and "[When I was] a child the dog was the only member of the family who could make life worth living." In other remarks, dog owners elaborated just what it was about the relationships with their dogs that made them preferable to human beings: always being there, always loving, and comparatively uncritical. In other words, the relationship with the animal—because it is based largely on the positive features perceived by the owners—manages to avoid those conditional and judgmental features that are so inconvenient in human relationships.

Archer also found "convincing evidence that people usually view their relationship with pets as similar to those they have with children"—for example, playing with their pets, talking to them in baby talk,* and cuddling them. Language directed toward babies and young children shows a number of specific characteristics that marks it out from the language used with adults. Such language is referred to as "motherese" and consists of a number of features, such as short utterances, with many imperatives and questions, repetitions, simple sentences, and tag questions (those ending with "aren't you?"). Kathryn Hirsh-Pasek and Rebecca Treiman examined recordings of dog owners talking to their dogs for such features in their speech, comparing the type of language spoken to the dog to that used in conversation with human babies. They found that nearly all the characteristics of motherese were present in these one-sided conversations with pet dogs, suggesting that a pattern of language used to aid interactions with young children has readily been co-opted for interacting with other social beings who are, like infants, presumed to be at a lower level of understanding than adult humans.[†]

The use of motherese is just one of the indications that the interactions people have with pets are modifications of those they have with other humans. Dogs and cats are mammals, like us, whose emotions and moods are similar to ours, although the ways they express them are different. Oskar Heinroth, one of the pioneers of the ethological tradition,[‡] described animals as "emotional people of extremely poor intelligence," a view shared by Archer: "He is right to the extent that it is the emotional similarity that people recognize in animals. This forms the basis of being able to communicate with them by visual and auditory signals, and by touch . . . and by sharing object play with them."

---

*This is true not only in the case of pet owners. When I take my cat to the vet, she talks to the cat, referring to me as "Daddy."
[†]This may not always be the reason for speaking motherese since it is also used in intimate adult relationships.
[‡]The tradition that one can only understand the characteristics of any species by observations made in the natural (as opposed to experimental) context.

]]]]] The Strength of Human Love for Pets

From time to time, reports appear in newspapers confirming the strength of devotion that some people bestow upon their pets. At a July 2005 wedding in Wanganui, New Zealand, the groom, Glen Armitage, designated his dog as best man, by no means the first reported case of a dog in this role.* There have been cases reported of people going one stage further and "marrying" their pet,† and there is now a Web site to make that process quicker and easier, as well as lucrative for the site owner. If you log on to www.marryyourpet.com, you will be able to choose between a "Simple Wedding" at only ten dollars, for which you can "marry your pet online and receive an official certificate of your happy day," in addition to which "all married couples can have their picture on the Marry Your Pet Happy Couples page,"‡ or a "Big Wedding" at eighty-five dollars, which brings the extra bonus of "an 'I married my pet' T-shirt so you can show the whole world just how much you cherish your pet."§ Or you could shell out two hundred dollars for the "Biggest Wedding," which gets you not only an online marriage, T-shirt, and certificate but also a "hand embroidered, personalized wall plaque to always remind you of your special day."

The marryyourpet.com site carries a disclaimer advising, inter alia, that "by marrying your pet he/she may be entitled to half your house and all your income," so you have been warned! But despite this warning, it is clear that a "marriage" to a pet has no legal significance or recognition. And it is not because of the *desire* for any form of legal recognition that some owners choose to make this gesture; rather it is

---

*At an August 2003 wedding in Settle, England, the couple's Alsatian, Barney, was dressed in a bow tie for the occasion, and it goes without saying that he accompanied the newlyweds on their honeymoon.
†For example, as reported in the "people" section of the *South Bend Tribune,* October 2, 1979, page 2, and in *Reader's Digest,* volume 116, February 1980, page 136.
‡These are prices as of late 2005.
§According to my calculation, that makes the cost of the T-shirt seventy-five dollars— a real bargain!

because they feel so much love for their pet that they want to affirm their commitment in a public way.

A more common example of a demonstration of love for a pet is seen when an owner offers a reward for finding a missing animal (and from time to time petnappers extract a ransom from a pet's loving owner). Another example is the far-from-rare occurrence of a deceased person's having bequeathed a substantial legacy to be used for the benefit of a pet, occasionally making the animal a millionaire. And sometimes in divorce cases a battle breaks out for the custody of a pet, a battle often conducted with a vehemence that other divorcing couples reserve for custody disputes over their children.

An even more widespread indication of the strength of people's attachment to their pets can be observed from the nature of pet owners' reactions to the loss of their animal, with the average length of the owner's bereavement following the death of a pet being some six to eight weeks. When a pet dies, the owner's feelings of grief are often very similar to those experienced due to the loss of a spouse or partner, a phenomenon first noted by researchers in the 1970s and 1980s. More recent studies by Elaine Drake-Hurst and Marilyn Gerwolls have also demonstrated parallels between the feelings of grief that follow the death of a human loved one and the grief prompted by the death of a pet, as has John Archer, who reported that a substantial number of pet owners surveyed in the United Kingdom, the United States, and Israel, were willing to admit that the death of a pet would make them cry.

While it seems clear from all this research that the *nature* of pet owners' grief is broadly similar to the grief suffered through the loss of a human loved one, it is less clear what *levels* of stress and depression are evoked by the grief from pet loss. Some studies have found these levels to be considerably lower than when a person is suffering the loss of a human loved one, while other research suggests that the levels of the grief are just as intense as those found after a human death. In yet another study, Mary Stewart investigated the effects of grief on pet owners due to the loss of their animal and found that as many as

18 percent of her survey group "were so disturbed that they were unable to carry on with their normal routine," and one-third of her subjects, although not quite so badly affected, nevertheless described themselves as "very distressed."

For many owners the only relationship in which they feel accepted and important is the one with their pet, and when that pet dies, much more is lost than the animal. The companionship, security, comfort, acceptance, love, and feelings of being needed and important—all are taken away with the pet's death, creating vacuums that explain why the death of a beloved pet can represent a profound loss. The closeness of owners' feelings for their pets was investigated by Sandra Barker and Randolph Barker, who found that dog owners generally felt as close to their pets as to the closest member of their family, and in one-third of cases the dog owners felt *closer* than to any human family member.

Another aspect of love for pets was investigated in an observational study conducted by Stephen Smith, which showed that women have stronger feelings of attachment to their (nonhuman) pets than do men, one of the reasons why I believe that many women will develop loving relationships with humanoid robots in the decades to come.

]]]]] Some Benefits of Owning Pets

The study of human-animal relationships is a relatively new field of psychological research that started attracting strong interest during the 1980s. A number of studies have indicated that it is not only emotional comfort and satisfaction that we can derive from our relationships with our pets but also therapeutic benefits, including improvements in people's health, happiness, and general well-being.* These effects result mostly from the lowering of the blood pressure and from the relaxation

*The earliest attempts to use animals for therapeutic purposes appear to predate this by almost two centuries. In 1792, William Tuke and several other Quakers in York, England, established a retreat where the mentally ill could be cared for much more humanely than was usual in those days. Tuke's idea was to provide farm animals for the patients to look after, believing that this activity would reduce the patients' aggressive instincts and improve their discipline and self-control.

response in humans caused by stroking and other forms of interaction with their cats and dogs.*

The emotional well-being brought on by pets can manifest itself in several different therapeutic forms. A pet can be a constant source of companionship, by providing love and by acting as a surrogate friend. And in the case of dogs, they can also act as parent substitutes, a role created as a result of the emotional security a dog brings to a household, performing a task that helps to relieve stress.

Emotional and Sociological Benefits of Pet Ownership

Alicia Stribling has found that the more contact people have with their pets, the happier they are.† One physiological reason for this is described by Johannes Odendaal and is related to six neurochemicals in the brain that help to reduce blood pressure. Odendaal found that when the dog owners in his experiment interacted with their pets, there was an increase in the production of these chemicals in the brain, including dopamine, phenethylamine, and endorphins, which are related to feelings of happiness and well-being, and at the same time there was a reduction of all the stress hormones, like cortisol.

Karen Allen has compared the relative benefits of having the social support of a friend or spouse with the therapeutic effects of a pet and found that a dog provides more effective social support for reducing stress than does a spouse! Two hundred forty married couples, of whom half were pet owners, were asked to perform two tasks known to induce stress: solving some problems in mental arithmetic and plunging a hand into ice water for two minutes. These experiments were carried out several times by the partners in each couple in various combinations: alone, with a pet or a friend, with their spouse, and with both their spouse and their pet or friend. Allen discovered that the pet owners exhibited much lower baseline heart rates and blood-pressure levels than did nonowners, commenting that "while

---

*What little research has been carried out with pets other than cats and dogs has been insufficient to demonstrate any comparable benefits.
†For the purposes of her experiment, she defined happiness as "a satisfaction with life and a general sense of well-being."

the idea of a pet as social support may appear to some as a peculiar notion, our participants' responses to stress, combined with their descriptions of the meaning of pets in their lives, suggest to us that social support can indeed cross species." And as for the social-support value of the spouses, Allen found that participants made the *most* errors in the mental-arithmetic problems when their spouses were present but their pets were absent. As a result she speculates that one reason pets appear to elicit such calm responses is that they encourage the positive-feeling states that social-support theorists have suggested may enhance a person's ability to handle stress. Furthermore, talking to dogs, in contrast to talking to one's spouse, has been found to be related to greater life satisfaction, greater marital satisfaction, and better physical and mental health.

Physical Health Benefits of Pet Ownership

One of the first researchers to recognize the *physical*-health benefits of pets was James Serpell, who investigated the therapeutic effects of giving non–pet owners a cat or a dog for periods ranging from six to ten months. He found that not only did the subjects' self-esteem improve while the animals were with them but their physical health did as well. This phenomenon had been suspected by Judith Siegel, who carried out a quantitative study on a sample of 938 patients enrolled in Medicare, finding that older people who own pets become less stressed by major adverse events in their lives and make fewer visits to the doctor than do non–pet owners.

Several other studies within various branches of medicine and care have similarly documented clear therapeutic benefits from pet ownership. Perhaps the most dramatic effect is that noted by Erika Friedmann and Sue Thomas, who found that heart patients who own pets are more likely to survive the year following a heart attack than are those who do not. Of 87 dog owners in Friedmann's study, only 1 died within a year of having a heart attack (1.1 percent) while of the 282 who did not own dogs, 19 died within that same period (6.7 percent), a ratio of six to one. These results support an earlier study led by Warwick Anderson at the Baker Medical Research Institute in Australia, which indicated that pet owners had lower cholesterol levels than did nonowners and were therefore less at risk of heart disease.

Friedmann's original results were questioned by some researchers, but she has verified them more than once. In 2003, for example, she and her colleagues at Brooklyn College of the City University of New York reported in the *American Journal of Cardiology* on a group of 102 patients who'd had a heart attack in the previous two years, including 35 patients who had owned a pet. Her team investigated the variability in the heart rates of these patients, a measure that indicates how well the heart is likely to handle stress. An increase in variability is linked to a lower risk of heart disease and death, and Friedmann's group found that the variability measures were higher in pet owners than in nonowners.

In 1994 the results of studies such as these prompted the largest of all surveys up to that time—the Australian People and Pets Survey— a national investigation, conducted by Bruce Headey, of more than a thousand people aged sixteen and over, some of whom were pet owners and some not. The aim of this survey was to quantify the extent to which the therapeutic benefits of pet ownership reduces the medical needs of the owners. Headey found that people who owned a cat and/or a dog required, on average, 5 percent less expenditure on treatments and medications than did nonowners, which in the case of Australia meant a cost saving of 1.8 billion Australian dollars* across the whole country. And within this group the differences between the number of doctor visits and the levels of medication required by dog owners who felt close to their dogs,† and the medical care of those who either were nonowners or who had a dog but did not feel close to it, were even more marked than that average figure of 5 percent. Another

---

*Equivalent to $1.2 billion U.S. at that time.
†In order to determine how close a pet owner feels to their pet, Headey employed a "closeness to pet" measure that averages the answers to four questions, in which the subjects were asked to say whether they agreed or disagreed with certain statements: (a) "I feel close to my pet"; (b) "When things go wrong, it is comforting to be with my pet"; (c) "Having a pet around when people visit me makes it easier to get into conversation and create a friendly atmosphere"; and (d) "I have sometimes got to know people and made friends through having pets."

important result of this survey was to confirm the significance of pets in the lives of people who live without partners—the single, the separated, the divorced, and the widowed—confirming that dogs can act as surrogate companions for those who lack a satisfactory network of human "social support."*

Following his work on the Australian People and Pets Survey, Headey collaborated with Markus Grabke on a similar study for Germany, comparing the data for a group of 10,000 respondents to a socioeconomic survey that had been repeated after a gap of five years. Within the survey group, those who had owned a pet for five years or more benefited the most, suggesting that it is the bond with the animal, rather than its mere proximity, that creates the feeling of well-being that positively affects the owner's health. This implication, that the therapeutic effects of a dog vary according to how well the patient has bonded with it, confirms the results of an earlier study at the University of Nebraska, which found that interacting with a dog with whom the patient has already formed a companion bond resulted in an 8-percent decrease in blood pressure, relative to interacting with a dog with whom the patient has not bonded. Subsequently Headey speculated:

> At a fundamental level, the benefits of pets appear linked to the human desire to be close to nature and other living creatures. The famous zoologist Edward O. Wilson has called the beliefs that humans need and benefit from closeness and companionship with other species "the biophilia hypothesis," which he postulates is based on an inherent, biologically based "predisposition to attend to, and affiliate with, like and lifelike processes."[3, 4] About 50 percent of adults and 70 percent of adolescents who own pets report that they confide in them. It is most unlikely that all this communication and companionship is wasted.[5]

*Social support is a network of family, friends, colleagues, and other acquaintances to whom one can turn, whether in times of crisis or simply for fun and entertainment.

]]]]] Comparing Relationships

Gail Melson has found considerable evidence that children aged three or even younger establish relationships with pets that provide emotional comfort in times of stress. She asserts that "the ties that children forge with their pets are often among the most significant bonds of childhood, as deeply affecting as those with parents, siblings and friends."[6] This form of comfort extends to school-age children, as indicated by the significant though widely varying percentages of children, aged from five to fourteen, the subjects of various studies, who said that they would turn to their pets when feeling sad, afraid, upset, or stressed. Children's feelings about their pets are typified by remarks such as, "My dog is very special to me. We have had it for seven years now. When I was little I used to go to her and pet her when I was depressed and crying. She seemed to understand. You could tell by the look in her eyes." Because of remarks such as these, Michael Robin and Robin ten Bensel were led to conclude, "As children get older the pet acquires many of the characteristics of the ideal mother: unconditional, devoted, attentive, loyal and nonverbal."[7] And the roles that pets play in a child's emotional development have been further investigated by Sandra Triebenbacher at East Carolina University, who found that almost all of the children she surveyed (89 percent or more) said that their pets were important members of the family, that they loved their pets very much, and that their pets also loved them very much.

The important benefits of pets described in this chapter have thus far been discussed without any mention of what the human-animal relationship is like for the pets. Humans and animals might well have completely different perceptions of their relationship, and it is known that animals generally prefer companions from within their own species to human companions. One might therefore expect that pets do not give their all to their human owners, in which case it is inevitable that robots will have the potential to be even better companions than animals are, because robots will be designed and programmed to *enjoy* their interactions with humans to the fullest and to behave accordingly.

One important indicator demonstrated by the human love for animals is that humans are able to form bonds of love with nonhumans. Anyone who maintains that it is unnatural for us to love robots, on the basis that humans can love only other humans, therefore faces the instant refutation of their argument. Our love for pet animals also provides support for our understanding of why it is that many people form strong emotional attachments to robot pets, the subject of chapter 3. The virtual pets of today, and earlier generations of robots, share with real pets one strong negative property that creates great similarities between human-pet relationships and human–virtual-pet relationships: It is not possible to carry on a sensible conversation with either. True, some robots can talk, using speech-synthesis technology, but their conversational abilities correspond at best to those of a two- to three-year-old toddler. In fact, the current level of speech recognition and understanding by robots, as well as this lack of conversational ability, makes them in some ways inferior as communicators to those animals whose owners "know" that their pet understands them and "talks" to them. This might be stretching the bounds of credibility too far for the liking of some readers, but one could argue in support of an extension to Alan Turing's thinking*—namely, if a pet owner believes that their animal understands them and "talks" to them, then we should accept that, for this particular pet owner, their animal does indeed communicate with them. And as Sherry Turkle notes, a similar tendency has been observed in some elderly people, who believe that a robot designed to be of therapeutic benefit to them is in a relationship with them, this because the robot makes eye contact or acts in some other way that is relationship-driven when seen in humans.

The fact that our love for our pets is understood by psychologists to be a form of attachment, the same phenomenon psychologists now accept as being the basis of romantic love, the same phenomenon that can have as its object computers or other artifacts, suggests that attachment permeates throughout the human-animal-artifact continuum. How has this attachment process with animals evolved? Archer

*With only a little license, Alan Turing's famous test for intelligence in a computer can be summarized thus: If a computer appears to be intelligent, then we should accept that it is intelligent.

believes that pets have evolved in ways that manipulate the human species through a number of features that make our interactions with them potentially rewarding for us, so that pets appear to treat their owners with love and affection. Cats and dogs behave in ways that are appealing to their human owners. Dogs show obvious signs of affection and attachment to their owners and are very attentive to them, while cats, although more independent, appear to like being stroked and petted.

]]]]] Why Do People Love Their Pets?

Many people believe that strong feelings directed toward a pet are an indication of an inadequacy in the person's relationships with humans. This judgment is often applied to a woman who lives by herself, has no children, and dotes on her dogs or cats. It can also be found in the comments of some psychiatrists about patients who show strong attachments to their pets. But there is a certain amount of convincing evidence that this view is wrong, evidence that people who have more secure attachments in their close relationships with other adults are the ones who are most strongly attached to their dogs. This is the opposite of what we would expect if strong attachment to a pet resulted from difficulties in forming relationships with adult humans.

Since reciprocity is one of the most significant factors in prompting feelings of romantic love,* it seems likely that the reciprocity demonstrated by pets—the purring of a cat and the nuzzling and tail wagging of a dog—similarly contributes to the strength of affection felt by an owner for a pet, and that reciprocity will likewise be a contributing factor in the growth of affection felt by an owner for a robot, when that robot demonstrates its virtual affection for its owner. A common example of reciprocity in dogs is seen when one of them is tethered to a lamppost while the owner goes into a shop. Next time you see this happen, watch that dog while the owner is in the shop. The dog will most likely remain fairly calm, perhaps trying to peer through the glass

*See the section "Attachment and Love," page 26.

into the shop to see the owner. But when the owner returns to collect the pet, the dog will usually go into paroxysms of excitement, the owner's absence, albeit for a short time, having made the dog's heart grow fonder. In their study of human emotions: *A General Theory of Love,* Thomas Lewis, Fari Amini, and Richard Lannon explain this reaction as being part of the attachment process between dog and owner:

> They spend time near each other and miss each other; they will read some of each other's emotional cues; each will find the presence of the other soothing and comforting; each will tune and regulate the psychology of the other. . . .[8]

Sherry Turkle at MIT was one of the first authoritative researchers to draw a parallel between man's relationship with animals and his relationship with computers:

> Before the computer, the animals, mortal though not sentient, seemed our nearest neighbors in the known universe. Computers, with their interactivity, their psychology, with whatever fragments of intelligence they have, now bid for this place.[9]

The human propensity for loving our pets thus informs our understanding of the emotional attraction to computers, to robot pets, and to humanoid robots. For those people who value their relationships with their pets more highly than their relationships with other humans, it would not be surprising if a virtual pet or a robot were to be regarded in the same vein, supplanting humans as the most natural objects of human affection. Where such people lead, others will surely follow, as the joys and benefits of relationships with robots become well publicized.

3 Emotional Relationships with Electronic Objects

A relationship with a computer can influence people's conception of themselves, their jobs, their relationships with other people, and with their ways
of thinking about social processes. It can be the basis for new aesthetic
values, new rituals, new philosophy, new cultural forms.

—Sherry Turkle[1]

▼ Sherry Turkle, an MIT professor of social studies and technology
and director of the institute's Initiative on Technology and Self, was the
first to publish extensively on the effects of computers on society—
what computers are doing to us—a subject in which her thoughtful and
groundbreaking 1984 book *The Second Self* has become a classic.* The
above quotation is taken from chapter 5, in which Turkle describes how
some of the early owners of home computers, some expert programmers, and some artificial-intelligence researchers took to them in a
novel way, forming some sort of relationship with their computer. These
were the earliest forms of the relationships that many owners nowadays
develop for their virtual pets.

A virtual pet is a computer representation of a model of pet behavior, incorporating software that allows owners to interact with their virtual pets. The computer might be a PC or game console that displays
images of the virtual pet on its screen; it might be a microprocessor-
based[†] product such as a mobile phone or a Tamagotchi, with a much
smaller display than a PC screen; or it could be a microprocessor-based
toy that looks like an animal or a robot. No matter what its embodiment
and appearance, the principle is the same: The virtual brain of the vir-

The Second Self was reprinted in 2005 in a "twentieth anniversary edition" with an
update section added.
[†]A microprocessor is a single computer chip that performs the "thinking" function.

tual pet is simulated by software in some sort of computing device. In summary, the core of a virtual pet is a computer of some sort plus some software. Relationships between humans and virtual pets are therefore an extended form of human-computer relationship, extended by the embodiment of the microprocessor in a petlike design, whether it be the design of a creature on a screen, as with the Tamagotchi, or the design of a doll or some form of petlike body that itself creates a measure of emotional appeal.

]]]]] Attachment and Relationships with Objects

In chapter 1 we touched upon the subject of attachment, discussing how the process of attachment in childhood extends into adulthood, sometimes manifesting itself as romantic love. Here we examine the process of attachment in more detail, as it pertains to computers and to virtual pets such as the Tamagotchi.

The process of attachment is closely related to another psychological phenomenon—transitional objects.* The young child becomes attached to an object such as a crib blanket (often spoken of as the child's "security blanket"), an article of clothing, or a soft toy. These are items that help the child to make the emotional transition from being wholly dependent on its mother and other caregivers toward being independent.

The significance of transitional objects was first recognized by the British pediatrician and psychoanalyst Donald Winnicott, whose 1951 essay "Transitional Objects and Transitional Phenomena" had an enormous impact on child psychology.† Winnicott argued that such attachments represent a developmental stage whereby infants make use of an object over which they have control, to deal with and move on from their early attachment to their mother, who is less under the infant's control than is the transitional object.

*Often called "security objects."
†Winnicott presented this essay at a meeting of the British Psycho-Analytical Society in 1951, but it was not published until 1953.

Subsequently other psychologists investigated and came to accept the notion that transitional phenomena extended past infancy, through adolescence, and into adult life. As Robert Young explains:

Having abandoned the blanket, doll or teddy, one can still attach similar significance to other objects with a less addictive intensity. The sensuous, comforting quality and the sense of something that is favorite and to which one turns when in danger of depressive anxiety applies to all sorts of special things. Everyone's list will be different, but these days Walkmans have this quality for many adolescents, as do portable computer games for pre-teens and computers for adult devotees, whether they be merely enthusiastic word processors or totally committed "hackers." The same can be said of mountain bikes, fancy roller skates, expensive trainers, certain fashions in clothes—Champion sweatshirts and sweatpants and Timberland shoes in the case of my children.[2]

And on the consequences of the comfort given by transitional phenomena, Young asserts that

they can become more real and intimate than human relations *per se*. One of the consequences of the fetishism of commodities is that the products of human hands appear as independent beings endowed with life and entering into relations both with one another and the human race. This arises not only from the commodity form but also from the formation of character in the image of the commodity.[3]

He further posits that the relationship between persons and things thus becomes transformed "so that my best friend is my Walkman or my personal computer." Sherry Turkle's explanation of the effect of early encounters with transitional objects is that they create a "highly charged intermediate space between the self and certain objects in later life."[4] She observes that not only do children *project* their fantasies and desires onto their transitional *inert* playthings, they also *engage with* their relational artifacts, their crying, talking *electronic* playthings.

Robert Pirsig's 1974 bestseller, *Zen and the Art of Motorcycle Maintenance,* expounded on the subject of intense relationships with technical objects and how such relationships can evoke philosophical musings. The latter-day version of Pirsig's motorcycling hero is the computer hacker,* many of whom boast on the Internet about their skills. Those who love the technology and the process of programming, mostly young men who are more than willing to stay up all night "hacking," share a fascination with the computer and an addiction to it. Out of this fascination, this addiction, springs a kind of love for the computer. Sherry Turkle has suggested that hackers' assertiveness of their skills is probably a symptom of a basic human need to credit their own place in society, their own favorite activities, with meaning. In *The Second Self,* in a chapter entitled "Hackers: Loving the Machine for Itself," Turkle describes hacking as

> a flight from relationship with people to relationship with the machine—a defensive maneuver more common to men than to women. The computer that is the partner in this relationship offers a particularly seductive refuge to someone who is having trouble dealing with people. It is active, reactive, it talks back. Many hackers first sought out a refuge during early adolescence, when other people, their feelings, their demands, seemed particularly frightening. They found a refuge in the computer and never moved beyond.

The hackers that Turkle describes here are at those at an extreme end of the social spectrum. Some programmers became hackers because of their love for the process of solving difficult problems, and the computer evolved to be the perfect tool for them because of the immediacy of the feedback it gave. Many of those who considered themselves to be

---

*A highly proficient and enthusiastic computer programmer—a "virtuoso programmer," to quote Sherry Turkle.

hackers during the 1960s, '70s, and '80s also had very active social lives that often integrated with their computing lives, providing a decent break from hacking. The extreme cases represented by the hackers interviewed by Turkle were simply the normal extremes of the personality spectrum, overlaid on the spectrum of those who love solving problems, of which mathematicians and chess grandmasters are other examples. Turkle's hackers, because of their extreme position on the social spectrum, were the first to exemplify the type of person who will be likely to embrace the ideas of love and sex with robots.

Turkle quotes one hacker who explained to her why, after he had "tried out" having girlfriends, he preferred to relate to computers:

> With social interactions you have to have confidence that the rest of the world will be nice to you. You can't control how the rest of the world is going to react to you. But with computers you are in complete control, the rest of the world cannot affect you.

And Turkle explains the role of the computer in providing relationships for those humans who have nowhere else (or no one else) to turn to, as being based on the computer's interactive capabilities:

> One can turn to the world of machines for relationship. . . . And the computer, reactive and interactive, offers companionship without the threat of human intimacy. . . . The interactivity of the computer may make him feel less alone, even as he spends more and more of his time programming alone.[5]

Norman Holland goes one step further, explaining why computer programming has been likened to sex:

> When programming, the computer addicts are working with an ideal partner who understands them fully. They feel toward their machines as toward a true friend. This friend will not withdraw if a mistake is made. This friend will try to be an ever-faithful helpmate. And this friend is male.[6]

But why do computers assume this role? The answer seems to lie in the process of attachment. Relationships that are attachment-based have been found to possess four characteristic features:

(a) An attachment figure, subconsciously associated with the infant's mother, takes on the role of "proximity maintenance," providing the comfort of always being there when its presence is needed, whether it be needed to bestow praise or to help dissipate feelings of fear.

(b) A corollary of this positive feature of attachment is the feeling of "separation distress" that occurs when an attachment relationship is disrupted, when the mother figure is absent.

(c) A further positive feature, closely related to proximity maintenance, is the role of the attachment figure as a "safe haven," allowing a person who is distressed to find contact (i.e., close proximity), assurance, and safety. This role is not one for which the attachment figure is uniquely suited, but just as an infant is more easily calmed by its mother than by another adult, so an adolescent or adult is normally calmed more easily by their attachment figure than by an alternative.

(d) An attachment figure has a role as a "secure base" from which to explore the world. A young child whose attachment figure is nearby and accessible will feel relatively comfortable in exploring strange and new environments but uncomfortable when lacking the proximity of their attachment figure. Similarly, an adult will normally feel more secure when exploring a new career opportunity or an unusual leisure activity if their romantic partner is accessible.

Without spreading the bounds of credulity too far, it isn't difficult to see how each of these four features can apply not only to human attachment figures but also to *artifacts* that serve the role of attachment figures, such as teddy bears, dolls, and computers. A young child likes to cuddle its doll or teddy bear (proximity); the child dislikes having its beloved toy taken away from it (separation distress); if the toy or doll is not actually within the child's reach, it is at least comforting for the child to know that it is nearby and accessible (providing a safe haven); and with the knowledge that the doll or bear is at hand or accessible, a young child will feel more confident about activities that involve exploration and discovery, activities that start from a "secure base." Replace "young child" with "adult," and replace "doll or teddy bear" with "computer" and any of you who are regular computer users will most likely be able to sympathize with the following rationalization: You like to interact with your computer because it responds to your input on the keyboard and with the mouse (proximity); you do not like being unable to access your computer (separation distress) because you rely on it to help you with certain tasks, such as checking your e-mail; if you are not actually using your computer, you feel more comfortable when it is near enough for you to access it when it is needed (a safe haven in the event of a storm of tasks); and you feel confident about playing a new game, deciding on the menu for a dinner party, or choosing a vacation destination, because you know that the computer is there to be asked (i.e., Google or some other search engine) if advice is needed. These are all symptoms of attachment.

Since the attachment process begins in infancy,* it is perhaps only natural that children generally exhibit stronger feelings of attachment for their computers than do adults. While young children bond with their blankets and toys, older children are bonding in large numbers with their computers. A MORI (Marketing & Opinion Research International) survey of children in Britain, conducted in December 2003, found that 45 percent of the children surveyed considered their computer to be a trusted friend, while 60 percent responded that they were extremely fond of their computer. The corresponding figures for adults were lower, at 33 percent and 28 percent respectively, but still a significant proportion of the population. Furthermore, 16 percent of adults and 13 percent of children aged eleven to sixteen responded that they often talk to their computer. And evidencing a general belief

*See also the section "Attachment and Love," page 26.

in the future of emotional relationships with computers, 34 percent of the adults surveyed and 37 percent of the children thought that by the year 2020 computers will be as important to them as are their own family and friends. This strength of the appeal of computers has been described by Cary Cooper, professor of organizational psychology and health at Lancaster University, as a "technological umbilical cord."

These findings would appear to indicate a shift in values in modern society, from a norm where the lives and well-being of family members are paramount to an entirely different scale of values, a scale on which a serious computer crash is deemed more important than the illness of a family member. Should we be so surprised that in some individuals and in some families such a different scale of values might exist? I think not. We have already seen, in the previous chapter, that some dog owners value their relationship with their pet more highly than their relationship with their spouse. So why should we not expect similar feelings to be expressed by some people for computers, and in the future for robots? Readers who are horrified at the fact that more than 30 percent of those surveyed held such opinions can take comfort from one very important factor that will to some extent at least militate in favor of the importance of the human family member relative to that of the computer or the robot: Humans are irreplaceable; computers and robots are replicable.* Hopefully, this factor will sustain a reasonable measure of balance in the minds of the majority.

In exploring the type of relationship that develops between humans and objects, it is important to understand exactly what we mean by "relationship" in this context. The contemporary view of relationships held by social psychologists is that the partners in a relationship are fundamentally interdependent—that is to say that a change in one of the relationship partners will bring about a change in the other. Mihaly Csikszentmihalyi and Eugene Rochberg-Halton have shown that our daily lives are influenced to quite a significant extent by man-

---

*The replicability of robots and one of its major implications are discussed in the section "Three Routes to Falling in Love with Robots," page 127.

made objects and that through these influences we establish a sense of connectedness with those objects.* A relationship with an object is one in which the experience we have with that object brings about a change in us, while what we do with that object will usually bring about a change in the object itself, even if it is a very small change, such as having experienced some wear through being used or simply having its location changed by being moved. The form of connectedness (for which read "attachment") that Csikszentmihalyi and Rochberg-Halton maintain we develop with objects is thus derived from the influence on our daily lives and on our identities brought about by our interactions with those objects. In the case of computers, "interaction" is certainly the operative word. Whereas our interaction with most objects is limited to what we do with the object, and is therefore a one-way street, our interactions with computers are two-way (or multiway) interactions, during the course of which what we do to the computer (typing on the keyboard, clicking the mouse, and thereby participating *with* the computer in whatever task it is accomplishing) is part of a genuinely interactive process.

A novel approach in the investigation of attachment to computers is expounded by John McCarthy[†] and Peter Wright in their delightful paper "The Enchantments of Technology." They argue that the attachment some people experience toward computers is one born of an enchantment with the technology. Each of us has the capacity to be enchanted by different things—some of us by a painting, some by a string quartet, some by the smile of a child, some by a motorbike. Just as different people can be enchanted by different things, so different things have the power to enchant different people, and technology is one of those things that has the power to enchant.

This view of enchantment as the basis of attachment to computers is due partly to the ideas of John Dewey, arguably the most influential thinker on education in the twentieth century. Dewey's 1934 book

*See also the section "Attachment and Love," page 26.
†For those readers with some interest in AI, this is not the John McCarthy from Stanford University who coined the term "artificial intelligence" in 1955 but the applied psychologist from University College, Cork.

Art as Experience asserts that experience is created by the relationship between a person and the tools that they use, the tools that form part of their environment. Dewey discusses a kind of sensual development of the relationship between a person and their environment, a development derived from a combination of the senses that familiarize the person with their environment. He uses as an example a mechanic working on an engine. When the mechanic is totally absorbed in his work, he sees, hears, smells, and touches the engine, and through these senses he diagnoses what is wrong. Being completely immersed in his work, totally focused on the task at hand, the mechanic develops a relationship with the engine. Because of his senses, he is caught up with what we might call the "personality" of the engine.

Another researcher who turned his attention to the enchantment of technology was the anthropologist Alfred Gell, who views the cause of this form of enchantment as being the power behind the enchanter. Gell suggests that the power of technology to enchant derives from our sense of wonder at the skill of that technology's creator—the process of creating the technology being more enchanting than the technology itself. But without any pleasure or similar emotions coming from the *experience* of a technology, McCarthy and Wright doubt the capacity of that technology to delight. To them enchantment also involves a sense of pleasure that is derived from the experience of novelty.* When your computer does something clever for the first time, something that satisfies you, there is a heightened feeling of pleasure. The satisfaction *contributes* to a state of enchantment, but it is the pleasure of novelty that turns satisfaction *into* enchantment. This is why working with computers and with software holds great potential for enchantment, because software is not always repetitive and boring—it often has the capacity to surprise, to create the unexpected. Consider, for example, a program designed to compose music. You might sit and listen to one new composition after another, with little to arouse your interest for a while. But then, out of the blue as it were, the program

---

*Bill Yeager points out that for masochists enchantment could involve a sense of pain and that many hackers fit into this category, because programming, solving tough problems, and fixing programming bugs are perceived by some as painful yet enjoyable.

produces a composition that you find very much to your liking. This new piece of music not only pleases you as a piece of music, but also surprises you by the power of the technology to compose, and it is in this surprise, and in the technology that creates this surprise, that the power of enchantment resides. This capacity to surprise is also evident to some extent in the behavior of virtual pets—an owner can rarely be certain how a virtual pet will behave.

McCarthy and Wright also explore the vicarious, visceral, and voyeuristic pleasure we derive from technology and how these forms of pleasure contribute to our feeling of enchantment. They use film as their example, but it could equally be computers or another form of technology. Transposing their exposition into a world with robots, the voyeur in us will experience the joy of watching the new and the wonderful, our vicarious nature will endear a robot to us if it repeatedly performs in novel and amazing ways that fit together as a coherent behavior pattern, while our visceral selves will rejoice in the thrill, joy, and apprehension of interacting with robots.*

]]]]] On Anthropomorphism

Some people regard anthropomorphism as a *belief,* even if only a transitory one, that the object in question is essentially human, a definition with which I disagree. In my view it is important to recognize the distinction between *believing* that a computer, for example, is essentially human and merely treating it in the same or similar ways as one might if it *were* human.

In the early days of artificial intelligence, Joseph Weizenbaum was perplexed by the extent to which people would anthropomorphize his computer program ELIZA:

*In line with his comment in the previous footnote, Yeager draws a parallel with the "dark side" of computing that attracts some people—those annoyances experienced when your computer hardware misbehaves, when your Internet connection fails for some reason, when pop-up advertisements appear when you least want them to, when your e-mail experience is harassed by spam or disrupted by a virus. These are irritations for most of us but provide a source of masochistic pleasure, thrill, joy, and apprehension for some.

I knew of course that people form all sorts of emotional bonds to machines, for example to musical instruments, motorcycles, and cars. And I knew from long experience that the strong emotional ties many programmers have to their computers are often formed after only short exposures to their machines. What I had not realized is that extremely short exposures to a relatively simple computer program could induce powerful delusional thinking in quite normal people.

What caused many of ELIZA's users to anthropomorphize was the humanlike snatches of conversation that the program generated:

USER: Men are all alike.
ELIZA: In what way?
USER: They're always bugging us about something or other.
ELIZA: Can you think of a specific example?
USER: Well, my boyfriend made me come here.
ELIZA: Your boyfriend made you come here.
USER: He says I'm depressed much of the time.
ELIZA: I'm sorry to hear you are depressed.[7]

It is easy to understand how someone could quickly begin to ascribe human characteristics to such a conversation partner.

The first grandmaster loss to a chess program led to an amusing example of anthropomorphism with computers. During a tournament in London in 1977, the English grandmaster Michael Stean was defeated by the program Chess 4.6, which had been developed at Northwestern University.* At one moment in the game, when the program had found a tactic that he'd overlooked, Stean referred to the computer as a "bloody iron monster," and some moves later he exclaimed, "This computer is a genius." Stean was not explicitly attributing humanlike qualities to the computer but employing the descriptive term "mon-

---

*This was a game of "speed" chess, in which each player has only a few minutes to make all their moves.

ster" as one might for a naughty child and "genius" as the ultimate intellectual compliment regardless of to whom (or to what) it is being paid. It was the program's remorseless performance in this intellectually difficult task that so impressed Stean, a world-class player, as to cause him to anthropomorphize.*

If you are a computer user, you will most likely have complained at some time or other that your computer refuses to work. In doing so you have attributed to your computer one of the characteristics of a living being, and you will have started to regard it as having some sort of relationship with you—a master-slave relationship in which you expect it to do your every bidding. The ease with which we slip into such a frame of mind has long been the subject of investigation by psychologists and anthropologists, but it is only relatively recently, with the advent of intelligent computers, that it has been recognized that the level of such relationships can rise to the point where, instead of being our slave, we think more in terms of the computer as a kind of friend.

In their book *The Media Equation,* Byron Reeves and Clifford Nass describe interaction with computers as being fundamentally a social tendency, but in their view it is not *consciously* anthropomorphic. They regard such interaction as automatic and subconscious, a view that stems from the general denial by most people that they treat computers as social entities. Yet despite this common denial, people do interact with computers according to normal human social conventions—by being polite, for example—and if a computer violates such a convention, it is usually regarded by its human operator as being deliberately offensive or obstructive, clearly an example of anthropomorphism. I believe that it matters not if the anthropomorphism of computers is subconscious. What I feel *is* important is the effect that anthropomorphism has on the emotional attachment felt *toward* a computer. It is the combination of attachment and anthropo-

---

*Bill Yeager makes the interesting observation that many of the remarks made about computers, such as Stean's, are inadvertent (i.e., subconscious) and knee-jerk reactions, and that eventually the dividing line between the human and robot species might become so fine that the idea of anthropomorphism, as it relates to robots, will disappear altogether.

morphism that, in my view, facilitates in us the creation of a human-computer relationship. As computers become increasingly accepted through the process of anthropomorphism, so will computer users come to treat them more like partners than work tools. For "computer" read "robot" and the mental leap is made—robots as partners.

But we are getting ahead of ourselves. Before we examine why humans develop relationships with computers, let us first explore in more detail Yeager's comment (see footnote on page 76), identifying the anthropomorphism of computers as a subconscious reaction.

Following the publication of *The Media Equation*, now widely regarded as a classic in the field of human-computer relationships, Reeves and Nass extended their experimental research in collaboration with Youngme Moon. Their studies investigated how people apply the rules of human social interaction in their interactions with computers. What their research results demonstrated was a marked difference between how people *say* they regard computers and how they *behave* toward computers. Their results are based on some of the thirty-five experimental studies they carried out, studies that re-created a broad range of social and natural experiences in which computers often took the place of one of the humans in the interaction.

Nass and Moon's paper, "Machines and Mindlessness: Social Responses to Computers,"* makes clear at the outset that:

> of the thousands of adults who have been involved in our studies, not a single participant has ever said that a computer should be understood in human terms or should be treated as a person.[8]

In the light of this unanimity, the actual behavior of these thousands paints a stark contrast, leading Nass and his group to conclude that there is clear evidence that people subconsciously treat comput-

---

*I find the use of the term "mindless" in their paper to be most unfortunate in the connotations of stupidity that it suggests. The authors adopt "mindlessness" from a 1992 paper by Ellen Langer, where "subconscious" would in my view be far more appropriate. Where I paraphrase extracts of Nass and Moon's fascinating paper, I have therefore replaced "mindless" with "subconscious."

ers as having personality and "apply social rules and expectations to computers." The experiments they carried out were mainly based on situations described in the literature of experimental psychology. The same social situations were replicated, as were the same experimental stimuli, but instead of monitoring a human-human social situation, the experimenters replaced one of the humans with a computer. Before you start to think that this replacement creates a completely different form of interaction, pause for a moment to consider some important similarities: (a) Humans communicate using words—so do computers; (b) humans are interactive in that they respond to a social situation based on all of their prior "inputs" from the person with whom they are interacting—computers are also interactive, in that the way they respond is based on their prior inputs from the user during that session (and possibly during earlier sessions as well, if the software has been programmed to learn); and (c) computers fill many roles that have traditionally been filled by humans. It is against this background of similarity, rather than a background of complete difference, that the results of these experiments should be interpreted.

One series of experiments carried out by Reeves, Nass, and Moon investigated whether computer users attribute gender to a computer.[*] Three stereotypical attitudes were investigated: (a) Dominant men are assertive and independent—positive attributes—while dominant women are pushy or bossy—negative attributes; (b) people are more likely to accept an evaluation of their own performance if it comes from a man rather than from a woman; and (c) people assume that men know more about certain topics, thought of as "masculine" topics, than do women, while women know more than men about certain "feminine" topics. The experiment, designed to test whether these stereotypical attitudes extend to "male" and "female" computers, employed programs that incorporated male and female recorded voices saying *exactly the same* things.

Each of the participants in the experiment went through sessions

---

[*]Use of the word "computer" here implies a combination of a computer and its software.

with three computers, each running a different program. There was a tutor program, a program that tested the participants on the topics taught by the tutor program, and finally a program that evaluated both the participants' test results and the teaching abilities of the tutor computer. Both groups, men and women, regarded the female-voiced evaluator as significantly less friendly than the male, supporting the stereotypical view that an evaluation by a man is more acceptable than *exactly the same* evaluation by a woman. In addition, both groups treated praise from the "male" computer more seriously than *exactly the same* praise from the "female" computer and believed the tutor computer to be *significantly more competent* after it had been praised by the "male" evaluator computer, compared to when it had been praised by the "female" evaluator. Finally, the "male" computer was perceived as being more informative than the "female" computer on the subject of computers (a "masculine" subject), while the "female" computer was considered to be the more informative when tutoring in love and relationships (a "feminine" topic).

The clear evidence from these experiments confirms that both men and women tend to carry over stereotypical views on human gender to their interactions with computers. Yet when they were questioned after the experiments, the participants uniformly agreed that there was no difference other than voice between the "male" and "female" computers and that it would be ludicrous to think of computers in gender stereotypes!

Another series of experiments was devoted to an investigation into whether people are polite to computers, as they are to other people. Research in social psychology has revealed that when someone is asked to comment on another person in a face-to-face social situation—for example "How do you like my new haircut?"—the resulting comments tend to be positively biased, even when the genuine evaluation might be negative. This is because people are inherently polite to other people. Nass and his team replicated this type of situation by having participants work with a computer on a task, then asking each participant to evaluate the computer's performance. These evaluations were conveyed by a participant in one of three ways: to the computer itself; to another computer, which the participant *knew* to be another computer but which was identical for all practical purposes to the computer being evaluated; and as a pencil-and-paper questionnaire. The evaluations presented by the participants to the collaborating computer itself were found to be significantly more positive than the evaluations presented to the second computer and to those on paper (both of which produced identical, and presumably truthful, responses). The clear conclusion here is that people are polite to computers, this despite a uniform denial by the participants that computers have feelings or that they deserve to be treated politely.

In yet another series of experiments, Nass's team investigated the psychological phenomenon of reciprocal self-disclosure. Research psychologists have confirmed something that is intuitively obvious— the general reluctance of people to talk about their innermost feelings to anyone other than their nearest and dearest. The one pronounced exception to this rule is that people *will* often disclose their secrets to strangers *if* the strangers first disclose secrets about themselves.[9] Does this reciprocity of self-disclosure apply to people who are in conversation with a computer? In the experiment designed to answer this question, the participants were interviewed by a computer on a variety of topics. Where there was no self-disclosure by the computers, the interview questions were asked in a different manner, without suggesting in any way that the computer had feelings and without the computer's referring to itself as "I." Typical of the differences between these questions was

What has been your biggest disappointment in life?

in which there is no self-disclosure, and

This computer has been configured to run at speeds of up to 266 MHz. But 90% of computer users don't use applications that require these speeds. So this computer rarely gets used to its full potential. What has been your biggest disappointment in life?

in which the computer's question is preceded by an explanation of one of its "disappointments." A less technically oriented example from the same experiment was:

Are you male or female?

and

This computer doesn't really have a gender. How about you: are you male or female?

The results demonstrated that that when the computer reciprocated, by first disclosing something about itself before asking the question, the participants' responses evidenced more intimacy, in terms of both the depth and the breadth of the participants' self-disclosure, than when the computer disclosed nothing about its virtual persona.[10] So once again the evidence points to a human tendency to relate to computers in much the same way as the same human would relate to other humans in comparable social situations.

The weight of the evidence found by Nass and his colleagues from these and other experiments* leads to the conclusion that people subconsciously employ the same social "rules" when interacting with computers as they do when interacting with other people. And this despite

the fact that the participants in our experiments were adult, experienced computer users. When debriefed, they insisted that they would never respond socially to a computer, and vehemently denied the specific behaviors they had in fact exhibited during the experiments.[11]

It seems perfectly reasonable to explain this phenomenon on the basis of a combination of attachment and anthropomorphism—more

*For example, the dominant/submissive computer experiment discussed in the section "Robot Personalities and Their Influence on Relationships," page 132.

the latter in these experiments, because the participants did not interact with the computers for long enough for attachment to become the dominant factor. Nass and his group disagree, basing their arguments on a subtle but importantly different definition of anthropomorphism from the customary one.* Instead they prefer to treat such behavior by computer users as ethopoeia, responding to an entity as though it were human while knowing that the entity does not warrant human treatment or attribution. I feel that the line between subconscious anthropomorphism (as I and many others use the word) and ethopoeia is too fine, if it exists at all, to cause us any concern in this discussion.

]]]]] The Development of Social Relationships with Computers

Computers are increasingly being regarded as our social partners, and with the evidence amassed by Nass and his group it is not difficult to understand why. In addition to the examples of their experimental research described above, Reeves and Nass have also discovered that people prefer interacting with computers that have identifiable personalities, more so when a computer's personality matches their own and especially when the user actually experiences the process of the computer's adapting its own personality and style of communication to be increasingly like that of the user.† Yet another supporting argument for the view of computers as social entities is the liking that people develop for computers that praise them, preferring these computers to ones that offer no such compliments.

One area in which social interaction between humans and computers is often evident is the realm of games. The history of game playing by computers is littered with evidence that many humans anthropomorphize when competing against a computer program—for

---

*They define the anthropomorphism of computers as a *belief* that computers are essentially human, a considerably stronger connection than that usually implied by the use of the word.

†Evidence for this phenomenon can be found in chapter 1, in the "Similarity" subsection of "Ten Causes of Falling in Love," page 38.

example, Michael Stean's exclamation "bloody iron monster" and his dubbing the computer "a genius."* In an experiment designed to investigate the manner in which human game players are emotionally stimulated by computers, two social psychologists, Karl Scheibe and Margaret Erwin, arranged for forty students to play five different computer matching games against a machine,† while a tape recorder was left running to record the students' comments. Almost all of the students referred to the computer as they might a human opponent, making comments such as, "It's just waiting for me to do it." Interestingly, the students' vocabulary employed for the machine often included the words "he," "you," and "it," but never "she."

While game playing is perhaps one of the most sociable activities in which computers can participate and demonstrate their sociability, the breadth of computer applications in which software can be socially responsive is almost limitless. One increasingly common reason for interacting with computer technology is the availability for purchase, via the Internet, of just about every type of product. When we buy something from an Internet shop, the owners of that shop want us to return to buy more, so customer loyalty and commitment are important to them. In order to engender such feelings in us, these shops often use software designed to learn more about us from our shopping habits, information that might be used at a later date to engage our interest and encourage us to buy. A relatively simple example of this can be seen in the way that the Amazon site operates. When I buy a book from Amazon, it remembers my purchase and tells me what other books the software believes might interest me. The software on the site knows‡ who else has bought the book I have just purchased, and it

---

*See the previous section.
†The games required the human subjects to make a binary choice at each move—for example, zero or one, heads or tails. The computer program would try to guess what choice was coming next. The humans tried to fool the computer program by varying their choosing strategy.
‡The software "knows" in the sense that you and I know something, by remembering. That the software can readily be perceived by us as knowing something is a prime example of how you and I anthropomorphize computers. I could have written, more precisely, "The software stores the knowledge that . . . ," but there is no need to be pedantic, since it is already generally accepted that computers "know" whatever it is

knows what other books those same people have purchased from Amazon, so it is able to deduce that I might have similar tastes to those other people and recommends to me the other books most often bought by that group. Translating this crude (but presumably effective) approach to a world with robots, when I ask my butler robot to bring me a glass of a particular chardonnay, it will remember, and in the future it might ask if I would like it to go to the wine store to buy a similar wine that it knows is on special offer. In this way my butler robot will endear itself to me, just as Amazon hopes to do. But relating to technology does not always bring its emotional rewards in the form of an interactive process, such as the way I might interact with my robot butler. We can love our Furby, but the Furby does not love us. We care about the Furby, but we do so without needing the relationship to become two-sided. In a sense this is analogous to sex with a prostitute—the needs of the client do not include the requirement that the prostitute love him.

Why, then, do some humans develop social relationships with their computers, and how will robots in future decades replicate the benefits of human-human relationships in their own relationships with humans? To help us answer this question, we should first consider exactly what emotional benefits human-human friendships provide and then determine whether these benefits might similarly be provided by computers.

In his book *Understanding Relationships,* Steve Duck has summarized the four key benefits of human friendships as:

(1) A sense of dependability, a bond that can be trusted to provide support for one of the partners when they need it.

A dramatic example of human trust in computers and dependability on them can be seen in the progress made during recent years in

_____
that is stored in their computer memories. I'm grateful to Bill Yeager for pointing out that I am as guilty as anyone of anthropomorphizing in this way.

the field of computer psychotherapists. For four decades researchers attempted, without very much success, to replicate in software the experience of psychotherapy encounters, replacing a human therapist with a computer. But then a team at King's College London, led by Judy Proudfoot, developed a successful therapy program called Beating the Blues, for dealing with anxiety and depression. Their most important finding was that computer therapy, using their software, reduced anxiety and depression in a sample of 170 patients "significantly and substantially," to levels that were barely above normal.

The relevance of this progress to the subject of human-computer emotional relationships derives from the nature of the patient-psychotherapist relationship. In making the initial decision to visit a therapist, and in deciding to continue with the course of therapy after the first few visits, a patient places great trust in the therapist. This trust encourages the patient to divulge personal and intimate confidences to the therapist and to take the therapist's advice on sensitive emotional and other intimate problems in their lives. The fact that patients willingly divulge the same confidences, and take the same advice, when interacting with a computer therapist demonstrates an inherent willingness to develop emotional relationships on a trusting and intimate level with computers.* Furthermore, as we saw in chapter 1, the act of divulging intimate confidences is one of the ingredients that can quickly turn a relationship into love.

(2) Emotional stability—reference points for opinions, beliefs and emotional responses.

---

*It would perhaps be useful to remind the reader that throughout this book, when discussing any aspect of human-computer interaction, I employ the word "computer" to mean the combination of the computer hardware (the box, screen, keyboard, and mouse) with whatever software it is running. Here, for example, what the user is actually trusting is the software with which the user is interacting. But because the user sees the computer, feels the keyboard and the mouse, and because it is the computer that displays and possibly speaks the output generated by the software, while the software itself is invisible, the user talks about their interaction as being with the computer rather than the computer-software combination.

Endowing a robot with opinions and beliefs is, at the simplest level, merely a matter of programming it with the necessary data, which could take a form such as this:

OPINION: The Red Sox will lose to the Yankees tomorrow.

EXPLANATION: Their top four players are ill with the flu. They have lost to the Yankees in the last seven games between them. The Yankees have recently purchased the two best players in the country.

And as software is developed that can argue a case logically—for use in robot lawyers, for example—it will become possible for robots to argue in defense of their opinions and beliefs by making use of such explanations.

Giving a robot the means to express appropriate emotional responses is a task that falls within the development of a software "emotion module." Robot emotions are discussed briefly in the section "Emotions in Humans and in Robots" in chapter 4, and more fully in *Robots Unlimited,** with the Oz emotion module, Juan Velasquez's Cathexis program, and the work of Cynthia Breazeal's group at MIT among the best-known examples created to date. Research and development in this field is burgeoning, within both the academic world and commercial robot manufacturers, and especially in Japan and the United States. I am convinced that by 2025 at the latest there will be artificial-emotion technologies that can not only simulate the full range of human emotions and their appropriate responses but also exhibit nonhuman emotions that are peculiar to robots. This will make it possible for robots to respond to some human emotions in interestingly different ways from those exhibited by humans, ways that some people will most likely find to be more appealing in some sense than the emotional responses they experience from humans.

(3) Providing physical support (doing favors), psychological support (showing appreciation of the other and letting the other per-

*Chapter 10.

son know that his or her opinion is valued), and emotional support (affection, attachment and intimacy).

Physical support from robots will be a question only of engineering, of designing and building robots to have the necessary physical capability to perform whatever task is being asked of them. If the favor consists of mowing the lawn or vacuuming the carpet, such robots are already on sale. As time goes on, more and more tasks will be undertaken by special-purpose robots, of which the lawn mower and vacuum cleaner are merely the first domestic examples. Eventually there will be not only a vast range of robots, each of which can perform its own specified task, but also robots that can operate these robots and others, making it possible for us to ask one robot to accomplish all manner of tasks simply by commanding the relevant special-purpose robots to do their own thing.

Psychological support from robots will most likely be provided by robot therapists, programmed with software akin to that employed in the program Beating the Blues.

Emotional support will be an ancillary by-product of a robot's emotion module, one for which artificial empathy will be a prerequisite. It has been shown that so long as a computer appears to be empathetic—understanding and responding to the user's expression of emotion and appropriate in the feedback it provides—it can engender significant behavioral effects in a user, similar to those that result from genuine human empathy. Empathy in robots will be achieved partly by measuring the user's psychophysiological responses, as described in the next chapter. By converting this empathy into emotional support, robots will be laying the foundations for behavior patterns that will enhance their relationships with their users.

(4) Providing reassurance about one's worth as a person.

Our friends contribute to our self-evaluation and self-esteem by giving us compliments and repeating to us the nice things that other people have said about us. Friends also raise our self-esteem by listening, asking our advice, and valuing our opinions. All of this will be accomplished by a robot's conversational module, backed by scripts and other conversational technologies that teach a robot how to talk in a reassuring manner.

In considering the potential of robots to provide these various benefits of friendship, Yeager asks whether it is likely or even inevitable that we should entertain some doubts in the backs of our minds—to what extent will people in the middle of this century be saying to themselves, "But this thing is still only a machine"? To what extent will those whose strongest friendships are all or mostly with robots miss the angst of human-to-human relationships? It is my belief that such doubts and feelings will by then have dissipated almost entirely, partly because robots will be so convincing in their appearance and behavior and partly because people who grow up in an era in which robots are even more commonplace than pet cats and dogs will relate to robots as people nowadays relate to their friends.

]]]]] Sustaining Social Relationships with Computers

Timothy Bickmore and Rosalind Picard have conducted an extensive review of the research into the social psychology of human-human relationships and human-human communication, research that is also relevant to human-computer relationships. They found that people use many different behaviors to establish and sustain relationships with one another and that most of these behaviors could be used by computer programs to manage their relationships with their users.

One of the key elements of relationships—an element that until recently at least has been missing from the software designed to create relationships between a computer and a human—is the importance of maintaining the interest, trust, and enjoyment of the human. Maintaining interest can be a side effect of doing everyday tasks together on a regular basis, the collaboration on these tasks acting as a bonding agent. Maintaining the trust in a relationship can be achieved by "metarelational communication"—talking about the relationship in order to establish the expectations of each partner and to ensure that all is well in the relationship. Other contributing factors to maintaining trust are: (*a*) confiding in one's partner as to one's innermost thoughts and feelings—this increases both trust and closeness; (*b*) emphasizing commonalities and deemphasizing differences—this behavior is associated with increasing solidarity and rapport with one's partner; and (*c*) "lexical entrainment"—using a partner's choice of words in conversation.

Maintaining the enjoyment of a relationship can also come in a variety of ways: (*a*) the use of humor, which makes computers appear more likable, competent, and cooperative than computers that lack humor; (*b*) talking about shared past experiences and the expectations of future togetherness, especially when making use of reference to mutual knowledge; and (*c*) "continuity behaviors" related to the time people are apart, talking about the time spent apart and using appropriate greetings and farewells. All these are important strategies in maintaining a sense of persistence in a relationship.

Conversation in general is also an important element of relationships and has formed one of the biggest challenges to the AI community, ever since Alan Turing proposed his famous test for intelligence in 1950.* Most human relationships develop during the course of face-to-face conversations, and even small talk, such as regular use of the greeting "Good morning," can influence the development of a conversation, since it has been found to increase the trust of some computer users.† And a lesson learned from the development of "expert systems" software is that another way for an intelligent computer to garner a user's trust is by explaining and justifying its beliefs, decisions, and conclusions during a conversation.‡

*See the section "Emotions in Humans and in Robots," page 118.
†This applies when talking to extroverts, but no effect has been found in conversations with introverts.
‡"Expert systems" as the name suggests, are computer-based systems that incorporate human expertise, usually in the form of the "rules" that human experts employ when making judgments and recommendations. It has been found that users of such systems place more trust in a system's decision-making capabilities if the system explains its thinking to the user by referring to or describing the rules it employed when making a particular decision.

It is not only what we say in conversation that affects people's reactions to us; how we speak is also important. The way we address someone will usually depend on the form of our relationship with them: "David" is friendly; "Mr. Levy" is less so. "Hello" is friendly; "Good morning" is less so. Thus the forms of language used in a computer application, even if they are only in menus or some other form of text, signal a certain set of relational expectations on the part of the user. And the tone of voice produced by a computer's speech synthesizer can also be an important factor in shaping the attitude of a user to that computer. The more frequently a computer matches the user in intonation, the higher the user rates the computer on measures of familiarity, such as comfortableness, friendliness, and perceived sympathy.

In summary it would appear that all of the emotional benefits we have considered here, deriving from human-human relationships, could also be provided by computers. Similarly, the behaviors we have discussed here, those necessary to endear one human being to another, appear already to be capable of simulation and in some cases *have* been simulated, using conversational and other techniques that are the subjects of research in the AI community.

]]]]] Virtual Pets—the Tamagotchi

In 1975 a fad for Pet Rocks was started in California by Gary Dahl, a salesman. Here was a pet that required no care, no food, no walking and yet gave its owner a few moments of pleasure. The idea spread like wildfire, and within a few weeks of the inception of the idea Dahl was selling rounded gray pebbles at the rate of ten thousand per day, together with a *Pet Rock Training Manual*—a step-by-step guide to having a happy relationship with your geological pet, including instructions for how to make it roll over and play dead and how to house-train it. "Place it on some old newspapers. The rock will never know what the paper is for and will require no further instruction."

In the light of the widespread enthusiasm for Dahl's completely inanimate, amorphous pets with which their owners could enjoy no real interaction, the advent and huge commercial success of the Tamagotchi* should have come as no great surprise. The idea for this product was conceived by a Japanese mother for her children, to counter their problem of being unable to own a real pet due to lack of space at home. Depending on which reports one believes, the number of Tamagotchis sold during its heyday varied between 12 million and 40 million.†

The Tamagotchi fits into the palm of the hand and is shaped like a flattened egg, with a small LCD‡ screen on which a simple graphical representation of the virtual pet is displayed. The idea is that the owner must care for the Tamagotchi in its virtual world, by pressing buttons to simulate the giving of food and drink, the playing of games, and other behaviors that are typical of a mother-child relationship, ensuring that the Tamagotchi will survive and thrive. When the Tamagotchi "wants" something, it sounds an electronic beep to alert its owner and indicates its particular needs at that moment by displaying appropriate icons on the LCD. If the Tamagotchi is neglected, it can "grow ill" and "die," often causing heartbreak to its owner. The creature's behavior patterns were programmed to change with time, in order to give the owners the sense that each Tamagotchi is unique and therefore provides a unique relationship for the owner, just as each pet animal and each human child are unique.

A remarkable aspect of the Tamagotchi's huge popularity is that it possesses hardly any elements of character or personality, its great attraction coming from its need for almost constant nurturing. It is this nurturing theme that engenders, in many Tamagotchi owners, a feeling of love for their virtual pet, an experience that can substitute for the experience of owning and caring for a real pet or even a human baby. In Japan the biggest group of Tamagotchi owners has been women in their twenties, most of whom purchased their toy because they craved the experience of nurturing. In the mother-child and other relationships

---

*The name is a diminutive form of the Japanese word *tamago* (egg) and is thus intended to convey the idea of a lovable egg.
†The heyday of the original Tamagotchi was the second half of 1997. A new version was launched in the summer of 2005.
‡Liquid crystal display.

between humans, the nurturer nurtures as a natural consequence of her love for the nurtured one, and of the object's need for her nurturing. In the human-Tamagotchi relationship, the same elements of a human relationship exist, but they act in the reverse direction—it is the need to nurture the virtual pet that engenders the emotion of love, not the love that impels the nurturing instinct. And this desire to nurture creates in many Tamagotchi owners what Sherry Turkle calls "the fantasy of reciprocation."[12] Tamagotchi owners also want their virtual pets to care about them in return.

This nurturing instinct is a significant feature in human-pet relationships as well. In referring to the role of a pet as a surrogate child in a childless relationship, or as an additional child for parents, Marvin Koller explains that yet another role of pets is to prolong the parenthood process for middle-aged and elderly parents whose children have flown the nest:

> The family pet *always* needs attention, and the pleasure it brings its keepers derives partly from the sustained dominance and importance of those who take care of it. The need to be needed is powerful, and parents whose children have grown up are gratified by this sustained dependence of their family pet over the years.[13]

The literature abounds with anecdotes about Japanese Tamagotchi owners who go to great lengths to preserve the life and well-being of their virtual pet—businessmen who postpone or cancel meetings so as to be able to feed their Tamagotchi and attend to its other essential needs at appropriate times, women drivers who are momentarily distracted in traffic while responding to the beeping of their needy electronic creature, a passenger who had boarded a flight but felt compelled to leave the aircraft prior to takeoff—and vowed never to fly with that airline again—because a flight attendant insisted she turn off her Tamagotchi, which the passenger felt was akin to killing it. Every example reflects the attitude of devoted Tamagotchi owners that their lovable egg is alive, and a logical corollary of this virtual life is that the Tamagotchi can virtually "die." When death occurs, the owners can arrange for the virtual "birth" of a new creature, and in addition many owners pay proper respect to their departed creature by logging on to a Web site that offers virtual cemeteries where the owners can post eulogies to their departed ones. The belief that their Tamagotchi had died is a further indication that the owner has somehow regarded it as having been alive.

It was not only in Japan that the Tamagotchi craze gave rise to important life decisions such as whether to miss a business meeting or to take one's eyes off the road while driving. In Israel an important religious question arose that depended for its answer on whether a Tamagotchi was deemed to be alive. Orthodox Jews are not permitted to do anything on the Sabbath that constitutes "work," and in the strictest of Orthodox households this includes such acts as switching on and off the lights and other electrical and electronic equipment, unless the act of work is necessary for *pikuach nefesh*—"the saving of souls," an act of life or death. The question therefore arose, is the pressing of the buttons on a Tamagotchi, an act carried out in order to sustain the Tamagotchi's virtual life, covered by the "saving of souls" exception? The position of Tamagotchi owners on this issue is clear, but the rabbinate in Israel took a different view—namely, that it is not a real soul being saved by pressing the buttons, and therefore interaction with a Tamagotchi is forbidden on the Sabbath. Despite this ruling, the very fact that the rabbinate had to make a decision on the Tamagotchi issue underlines the widespread feeling that the Tamagotchi is alive and has a right to life.

The effect of the Tamagotchi and Furby crazes has been to spawn a culture in which electronic products are accepted as having lifelike properties. Sherry Turkle describes how children have been affected by this realization of some sort of life in man-made objects:

A generation of children is growing up who grant new capacities and privileges to the machine world on the basis of its animation. Today's children endow the category of made objects with properties such as having intentions and ideas. These were things previously reserved for living beings. Children come up with the new category "sort of alive" for describing computational animation, and they are increasingly softening the boundaries between artifact and flesh, as well as blurring boundaries between the physical real and simulation.[14]

But even though Turkle, when researching for her 1984 book *The Second Self,* grew to expect that children "might come to take the intelligence of artifacts for granted, to understand how they were created, and be gradually less inclined to give them importance," she was surprised at "how quickly robotic creatures that presented themselves as having both feelings and needs would enter mainstream American culture," remarking that "by the mid-1990s, as emotional machines, people were not alone."[15]

Turkle explains that as a result of this change in perception as to the aliveness of artifacts, "people are learning to interact with computers through conversation and gesture. People are learning that to relate successfully to a computer you have to assess its emotional state; . . . you take the machine at interface value, much as you would another person." And she discovered that in some people this change in perception can lead to a preference for interacting with an artificial creature rather than a real one, quoting children who, on seeing a pair of Galápagos turtles at the American Museum of Natural History in Boston, remarked that robot turtles would have been just as good, cleaner, and would have saved transporting the real ones thousands of miles. Turkle also observes that "when Animal Kingdom opened in Orlando, populated by 'real'—that is, biological—animals, its first visitors complained that they were not as 'realistic' as the animatronic creatures in the other parts of Disney World. The robotic crocodiles slapped their tails, rolled their eyes—in sum, displayed archetypal 'crocodile' behavior. The biological crocodiles, like the Galapagos turtle, pretty much kept to themselves."[16]

The relationship between Tamagotchi owners and their virtual pet has been compared to "parasocial" relationships. The term "parasocial" was coined by Donald Horton and Richard Wohl to represent the type of interaction that TV viewers have in mind when they imagine themselves becoming closely acquainted with the personalities of characters on their favorite shows: "After watching a television series for a period of time, viewers come to feel that they know the characters as well as friends or neighbors."

It has been found that the process of developing parasocial relationships bears many similarities to the process of developing real-life relationships. But Linda Renée-Bloch and Dafna Lemish assert that the development of an owner-Tamagotchi relationship is quite different from a parasocial relationship because, in the case of the Tamagotchi, it is not a human (TV) personality with which the relationship is developed, but the personification of a machine. They support their assertion with the argument that in the Tamagotchi relationship the owners can affect the life of the creature by their actions: "The very existence of the virtual partner to the interaction depends on responding to its demands." I take the opposite view. I hold that precisely *because* the owner can affect the virtual life of the Tamagotchi, the relationship is an even stronger form of parasocial interaction than that between a TV viewer and a favorite character, the dream of having an intimate closeness with that character being better realized in the case of the Tamagotchi because its owner controls, and has the power to enhance, the creature's virtual life—just as a human has the power to enhance and to some extent control (or at least affect) the lives of friends and loved ones. This type of power can already be seen in some interactive TV systems that allow a viewer to determine what happens next in a story line—should she kiss him passionately, slap his face, or run out of the room crying? Such systems enhance TV viewers' parasocial-relationship experience by adding the element of control, allowing them to gain an increased level of intimacy with the TV character in a similar way to how Tamagotchi owners relate to their virtual pet.

]]]]] Virtual Pets That Live
       on the Screen

Handheld virtual pets such as the Tamagotchi are the simplest form of the genre, based on low-cost electronics that allow a retail price of fifteen dollars or less. The next step up in complexity is the virtual pet that "lives" on the TV or computer screen, usually a cartoonlike character. The most believable and lifelike of these characters exhibit a variety of social cues: intelligence, individuality, sociability, variability, coherence, and some conversational ability. Add the ability to recognize the user's emotional state and other social cues and they will become utterly compelling.

Sherry Turkle notes that the behavior of a character in a computer game impels some computer users to anthropomorphize not only the virtual character but also the computer itself. This is hardly surprising, given that computer users often anthropomorphize their computer even when the task it is executing is not one involving any virtual characters. When a believable character appears on the screen, the tendency to anthropomorphize must surely be greater.

A popular example of a screen-based character that encourages anthropomorphism is the virtual girlfriend. A character of this sort was first announced in a 1994 advertisement in *PC Magazine*:*

Now You Can Have Your Own GIRLFRIEND
 . . . a sensuous woman living in your computer!
GIRLFRIEND is the first VIRTUAL WOMAN. You can watch her, talk to her, ask her questions and relate with her. Over 100 actual VGA photographs allow you to see your girlfriend as you ask her to wear different outfits, and guide her into different sexual activities. As a true artificial intelligence program, GIRLFRIEND starts with a 3000 word vocabulary and actually GROWS the more you use it. She will remember your name, your birthday, and your likes and dislikes. GIRLFRIEND comes with the base software [*sic*] and GIRLFRIEND LISA. Additional girls will be added. This program requires 7–10 MB of free space.

This type of character has recently been metamorphosed to create a new twist on the Tamagotchi concept. Rather than the user's lavishing

*Volume 13, page 483.

care on the virtual character as the path to giving her a long and happy life, the key with this virtual girlfriend, launched by the Hong Kong company Artificial Life in the autumn of 2004, is much simpler. It is money. For a monthly fee of six dollars (real money, not virtual dollars), customers can download an image of "Vivienne," a slim, talking brunette, to their cell phones and then spend much more (real) money sending her virtual flowers, virtual chocolates, and other virtual gifts, not to mention the essential spending on the cell-phone calls necessary to interact with Vivienne. In return for their generosity, customers are made privy to different aspects of Vivienne's life, such as meeting her virtual female friends, who also appear as images on the display screen of the cell phone. But if a customer neglects Vivienne, she refuses to speak.

Vivienne was followed in January 2006 by a virtual boyfriend for women, with other characters being planned by Artificial Life to cater to gay and lesbian customers.

]]]]] Robotic Virtual Pets

The highest form of virtual pet is one that moves around your room—for example, Sony's AIBO, a robot dog. AIBO's design was based on the ethology* of canine behavior patterns, and in particular on the research conducted by John Scott and John Fuller, and also that of Michael Fox. This body of research has provided a comprehensive categorization of canine behavior patterns that covers the whole range of a dog's activities and forms the basis for the AIBO's own behavior patterns, which include expressions of anger, disgust, fear, happiness, sadness, and surprise.

AIBO comes with a number of preprogrammed behavior patterns that encourage owners to project humanlike attributes onto their virtual pets. The AIBO plays, it sleeps, it wags its tail, it simulates feelings of affection and unhappiness. Sony describes the AIBO as "a true companion with real emotions and instinct."[†] Not everyone will embrace

---

*Ethology is a study of animals in their natural surroundings.
[†]At www.aibo.com.

this concept, but to a large extent any argument over this point is not of great import. What *is* important is that many people, especially children and the elderly, have been found by psychologists to behave with AIBO in the same way they would interact with real animals. And as the technology improves and robot pets become increasingly lifelike, the boundary between people's perceptions of robotic pets and their perceptions of real animals will become increasingly blurred.

As a result of its animal-like behavior, AIBO engenders feelings of love in many of its owners similar to those felt by the owners of real pets. Children's interactions with AIBO were investigated in a comparative study of seven- to fifteen-year-olds, which compared their AIBO interactions to their interactions with a real Australian shepherd dog. The majority of children in this study treated AIBO in ways one would treat a dog. As one child said, when asked how she would play with AIBO, "I would like to play with him and his ball and just give him lots of attention and let him know he's a good dog." Fifty-six percent of those surveyed by Gail Melson believed that AIBO had mental states (for example, feeling scared), 70 percent said that AIBO had personality, and 76 percent asserted that AIBO had moral standing (i.e., it could be held morally responsible or blameworthy for its actions and could have rights and deserve respect). Given how rudimentary AIBO is in terms of its capabilities, it is remarkable that so many children treated it not only as if it were a social agent (the focus of research by Reeves and Nass, albeit human, not dog) but also as having mental states and moral standing. It is therefore reasonable to conclude that as robots become increasingly lifelike in their behavior and as these children influence the adults around them and grow into adults themselves, more and more people will treat robots as if they are mental, social, and moral beings—thus raising the perception of robotic creatures toward the level of biological creatures.

The extent of the love of AIBOs demonstrated by their adult owners can be seen from the many AIBO Internet chat sites that testify to just how widespread these feelings of love are. In a study based on more than three thousand spontaneous Internet postings on AIBO discussion forums, a team led by Peter Kahn found that 42 percent of

EMOTIONAL RELATIONSHIPS WITH ELECTRONIC OBJECTS

forum members spoke of their AIBOs as having intentions or engaging in intentional behavior. For example, "He [AIBO] also likes to wander around the apartment and play with his pink ball or entertain or just lay down and hang out." Or, "He is quite happily praising himself these days." Some members (38 percent) spoke of AIBO as having feelings: "My dog [AIBO] would get angry when my boyfriend would talk to him," or "Twice this week I have had to put Leo [AIBO] to bed with his little pink teddy and he was woken in the night very sad and distressed." Some members (39 percent) spoke of AIBO as being capable of being raised, developing and maturing—for example, "I want to raise AIBO as best as I possibly can." Some (20 percent) spoke of AIBO as having unique mental qualities or personality, and 14 percent of the members of the forum imbued AIBO with a substantial measure of animism—for example, "I know it sounds silly, but you stop seeing AIBO as a piece of hardware and you start seeing him as a unique 'life-form,'" or "He seems so alive to me."

Kahn and his team raise this question: "What are the larger psychological and societal implications as robotic animals become increasingly sophisticated, and people interact less with real animals and more with their robotic counterparts? Our results provide some empirical data to begin to think about such a question. We are not saying that AIBO owners believe literally that AIBO is alive, but rather that AIBO evokes feelings *as if* AIBO were alive." Based on the research of Batya Friedman and her colleagues, it seems that these feelings arise because people actually *want* to perceive their AIBOs as real pets, and therefore they attribute doglike emotions to AIBO. The design of the AIBO has not yet been developed to the point where it can *have* simulated doglike emotions and express them in ways that its owner can appreciate, but such capabilities in robot pets will come, and they will probably not be long in coming. The relative successes in emotional modeling that have been built upon the findings of the ethology literature will undoubtedly lead to an increase in the study of ethology for this specific purpose, and when it is fully understood what makes dogs tick, it will be possible to develop increasingly sophisticated simulations of their emotional makeup and to employ such simulations in future artificial canines.

One crucial aspect of life and bonding that has not yet begun to be deeply explored by the developers of robotic pets and partner robots is aging. This is important not only because of the inevitability of our own eventual deterioration and death but also because of the learning processes and greater strength of bonding that can take place as we age. The depth and richness of behavior patterns in animals, including humans, is founded on the learning process and all that goes with it. As we get to know someone better with time, our relationship and intimacy with them can develop, grow stronger.

But the aging process in humans has a downside that will not necessarily be designed into robots—the inevitability of death. In theory at least, there is no reason robots will need to "die," and even if a robot suffers damage, it can be replicated, both physically (new body, same appearance) and mentally (a copy of the contents and intellectual capacities of its "brain"). The possibility therefore exists that while simulating the process of growing older alongside its owner, with all the benefits of greater bonding and greater intimacy that that will bring, robots will be able to continue to develop in this way but without ever dying. In the case of humans, impermanence is built in. In robots, impermanence can be built out, allowing them to continue to develop even after their human has passed away. This suggests fascinating possibilities, such as robots' being able to "outlove" their human partners, loving them more, and in better ways, than their humans love them. If the robot's brain has already absorbed everything it learned about its human from a previous long-term relationship, the robot might have a greater capacity for love and a greater knowledge about how to love than when it was first programmed.

]]]]] The Benefits of Forming Attachments to Robot Pets

The development of AIBO and other technological substitutes for pets has been inspired in part by the benefits that are known to derive from conventional human-pet relationships, and it is now known that there are also psychological and other benefits, especially for children and the elderly, in forming attachments with sociable robots. As we have seen in chapter 2, research into the therapeutic benefits of owning *real* pets suggests that simulated pets might bring therapeutic benefits to the elderly, to the disabled, and to emotionally disturbed children, as the real-world consequences of the users' treatment of their *virtual* pets are also simulated by the virtual pets' behavior patterns.

The use of robot pets as companions and carers for the elderly is a research topic that is gathering great momentum, particularly in Japan and the United States, and partly because feeling cared for is known to have profound effects on a patient's physiology, cognition, and emotional state. Governments are now worrying about how their countries' social services will be able to cope with huge populations of senior citizens. The U.S. Census Bureau, for example, has estimated that the elderly population in the United States will more than double between 2005 and 2050, to 80 million people. How will the elderly be provided with the emotional and physical care they need?

A research team led by Nancy Edwards at Purdue University is investigating the use of robots as a possible solution, providing a simulation of caring that is expressed partly through the content of a robot's speech; partly through its voice, tone, and the timing of its speech; and also through the use of appropriate facial expressions and postures. Human communicative behaviors that could be employed by a robot to elicit the perception of feeling cared for include demonstrations of empathy and comforting behavior, both of which are within the grasp of current AI research. And facial expressiveness by physical therapists (smiling, nodding, and frowning) has been found to be significantly correlated with short- and long-term functioning in their geriatric patients.

Edwards and her team base their idea of using robot pets as carers on the known therapeutic benefits of real animals for the elderly:

Hundreds of clinical reports show that when animals enter the lives of aged patients with chronic brain syndrome (which follows from either Alzheimer's disease or arteriosclerosis) that the patients smile and laugh more, and become less hostile to their caretakers and more socially communicative. Other studies have shown that in a nursing home or residential care center, a pet can serve as a catalyst for communication among residents who are withdrawn, and provide opportunities (petting, talking, walking) for physical and occupational rehabilitation and recreational therapy. Thus, is it possible that robotic pets—such as Sony's robotic dog AIBO—can provide the elderly with some of the physiological, cognitive, and emotional benefits of live pets?[17]

Solid evidence that computers have the capacity to instill a sense of caring was revealed in a study carried out by Timothy Bickmore as part of his Ph.D. research at MIT. Bickmore employed an animated, talking character named Laura, a virtual fitness consultant, whose screen image showed her with bobbed chestnut brown hair. Laura was designed to advise users on how to improve their training regimes, and the participants in Bickmore's experiment interacted with Laura for ten minutes every day for a month, answering her questions about their workouts and being guided by her advice on how to overcome various obstacles they encountered in doing their daily exercise. Two versions of Laura were employed for the experiment, with roughly half the participants interacting with a version that incorporated a full range of caring behaviors that included providing health information, giving feedback on the participants' exercise behavior, and encouraging them to commit to exercise. This "caring" version would sympathize with any participant who claimed not to feel well enough to exercise that day, the sympathy including suitable facial gestures as well as an appropriately sad tone of voice. The other group of participants interacted with a version of Laura that provided the same health advice but none of the caring interactions.

The result after one month was dramatic. Those participants who had interacted with the caring version of Laura exhibited a significantly greater agreement with four statements about their experience than did those who worked with the noncaring version: (a) "I feel that Laura cares about me in her own unique way, even when I do things that she does not approve of." (b) "I feel that Laura, in her own unique way, is genuinely concerned about my welfare." (c) "I feel that Laura, in her own unique way, likes me." (d) "Laura and I trust one another." When the participants were asked at the end of the month whether they would like to continue working with Laura, those who had interacted with the caring version responded much more positively than those in the other group, and significantly more participants (69 percent) in the "caring Laura" group chose to sign off their final session with "Take care Laura, I'll miss you," rather than with the proffered alternative of simply "Bye"—whereas in the "noncaring Laura" group only 35 percent chose the more sentimental sign-off option.

Bickford's results indicate that a suitably programmed virtual character can significantly increase a user's perception of being cared for, even when the user is a very bright, computer-savvy student who knows that computers do not genuinely care for their users.

When robot pets are made sufficiently lifelike, with warm bodies, soft artificial flesh, and perhaps with artificial fur, their owners will most likely derive even greater therapeutic benefits than the owners of real pets get from stroking them and from other forms of interaction with them, given that robot pets will also be able to carry on some sort of meaningful conversation, however rudimentary it might be. For children the social benefits of such attachments would include the learning of decent social behavior—being kind to their virtual pets—and unlearning negative social behavior.

]]]]] From Virtual Pet to Humanoid Robot

The transition from relating to a simple battery-operated toy animal to relating to video-game characters, then to computer characters, to robot animals, and finally to human-looking robots is not a difficult one to make. Given that children have already been shown to form emotional attachments to virtual and robotic pets and that, at the opposite end of the age spectrum, the elderly are showing a similar tendency toward carer robots, it seems extremely likely that this phenomenon will eventually extend to all generations, when today's children, who grow up loving their robot pets, have turned into tomorrow's adults.

And by adding intelligence to a robot and making the entity convincingly humanlike rather than doglike in appearance, robot manufacturers will enhance the user experience to such a great extent that the adult who twenty years earlier would happily play with a simple robot pet will be likely to enjoy the company of one of its successors—the humanoid robot. Cynthia Breazeal, who led the design of the sociable Kismet robot at MIT, found that when she finished her Ph.D. and had to leave Kismet behind in the robot laboratory, she suffered withdrawal symptoms and described a sharp sense of loss. "Breazeal experienced what might be called a maternal connection to Kismet; she certainly describes a sense of connection with it as more than with a 'mere' machine."[18]

Those who will adapt the best to the era of life with robot friends, companions, and lovers, will most likely be those who grew up surrounded by other forms of robot, including possibly a robot nanny. The research currently under way on a major scale, particularly in Japan, which is aimed at developing carer robots for the elderly, will have as one of its spin-offs carer robots for children, from infants upward. It is only natural that a child who grows up in a house with a robot nanny—particularly if the nanny was kind to the child and loved by it—would be highly receptive, as it developed toward adulthood, to the concept of friendship and love with other types of robot.

> > > > > > > > > > >

4 | Falling in Love with Virtual People (Humanoid Robots)

A sociable robot is able to communicate and interact with us, understand and even relate to us, in a personal way. It should be able to understand itself and us in social terms. We, in turn, should be able to understand it in the same social terms—to be able to relate to it and to empathize with it. Such a robot must be able to adapt and learn throughout its lifetime, incorporating shared experiences with other individuals into its understanding of self, of others, and of the relationships they share. In short, a sociable robot is socially intelligent in a humanlike way, and interacting with it is like interacting with another person. At the pinnacle of achievement, they could befriend us, as we could them.

—Cynthia Breazeal[1]

]]]]] Attitudes to Relationships

▼ It is well established that people love people and people love pets, and nowadays it is relatively commonplace for people to develop strong emotional attachments to their virtual pets, including robot pets. So why should anyone be surprised if and when people form similarly strong attachments to virtual people, to robot people? In response to this question, some might ask, "But why would anyone want to?" There are many reasons, including the novelty and the excitement of the experience, the wish to have a willing lover available whenever desired, a possible replacement for a lost mate—a partner who dumped us. And psychiatrists will no doubt prescribe the use of robots to assist their patients in the recovery process—after a relationship breakup, for example—since such robots could be well trained for the task, providing live-in therapy, including sexual relations, and benefits that will certainly exceed those from Prozac and similar drugs.

I believe that one of the most widespread reasons humans will develop strong emotional attachments to robots is the natural desire to have more close friends, to experience more affection, more love. Timothy Bickmore explored the concept and implications of having computer-based intimate friendships in his 1998 paper "Friendship and Intimacy in the Digital Age," in which he surveyed the state of friendship in our society and found it to be "in trouble." Bickmore explains:

> Many people, and men in particular, would say they are too busy for friends, given the increasing demands of work, commuting, consumerism, child care, second jobs, and compulsive commitments to television and physical fitness.

Bickmore supports this assertion by quoting from the 1985 *McGill Report on Male Intimacy*:

> To say that men have no intimate friends seems on the surface too harsh, and it raises quick observations from most men. But the data indicate that it is not very far from the truth. Even the most intimate of friendships (of which there are very few) rarely approach the depth of disclosure a woman commonly has with many other women. Men do not value friendship. Their relationships with other men are superficial, even shallow.[2]

Bickmore also quotes the statistic that "most Americans (70 percent) say they have many acquaintances but few close friends," and he then posits that "technology may provide a solution." His argument is clear and convincing. Given the great commercial success of the rather simple technology employed in virtual pets such as the Tamagotchi and the AIBO robotic dog, and the popularity of the even simpler conversational technology employed in ELIZA and other "chatterbot" programs,*

---

*Chatterbot (or chatbot) is the generic name of the ELIZA-like programs that can carry on a conversation, appearing always to understand the user's previous utterance while in fact understanding absolutely nothing.

it seems clear that a combination of these technologies, with additional features for self-disclosure and simulating an empathetic personality in the robot, would provide a solid basis for a robotic virtual friend. It is of course reasonable to question why someone would have time for a robot friend but insufficient time for a human one. I believe that among the principal reasons will be the *certainty* that one's robot friend will behave in ways that one finds empathetic, always being loyal and having a combination of social, emotional, and intellectual skills that far exceeds the characteristics likely to be found in a human friend.

AIBO is clearly the most advanced virtual pet to make any commercial impact thus far, but AIBO's vision and speech capabilities are limited in comparison with the best that technology could offer today if cost were no object. Nevertheless, even with these limited capabilities, AIBO appeals to many children and adults as a social entity. Progress in creating everyday lifelike behavior patterns in robots will increase our appreciation for them, and as robotic pets and humanoid robots increasingly exhibit caring and affectionate attitudes toward humans, the effect of such attitudes will be to increase our liking for the robots. Humans long for affection and tend to be affectionate to those who offer it.

As a prerequisite of adapting to the personality of a human, robots will need to have the capacity for empathy—the ability to imagine oneself in another person's situation, thereby gaining a better understanding of that person's beliefs, emotions, and desires. Without empathy a satisfactory level of communication and social interaction with others is at best difficult to achieve. For a robot to develop empathy for a human being, it seems likely that the robot will need to observe that person's behavior in different situations, then make intelligent guesses as to what is going on in that person's mind in a given situation, in order to predict subsequent behavior. The acquisition of empathy is therefore essentially a learning task—relatively easy to implement in robots.

The psychological effect on computer users of interacting with an empathetic program was evaluated in an experimental study at Stanford University. The participants were asked to play casino blackjack on a Web site, in the virtual company of a computer character who was represented by a photograph of a human face. The computer character would communicate with the participants by displaying text in a speech bubble adjacent to its photograph. The participant and the computer character "sat" next to each other at the blackjack table, and both played against an invisible dealer. After each hand was completed, the computer character would react with an observation about its own performance and an observation about the participant's performance.

Two versions of the program were used, one in which the computer character appeared to be self-centered and one where it appeared to be empathetic. In order to simulate self-centeredness, the character would express a positive emotion if it won the hand, by its facial expression and what it said, and a negative emotion if it lost, but it showed no interest in whether the user won or lost. The empathetic version displayed positive emotions when the participant won a hand and negative emotions when the participant lost.

The investigators found that when the computer character adopted a purely self-centered attitude, it had little or no effect on the participants' reactions to its virtual personality. But when the computer character appeared to empathize with the users' results at the blackjack table, the participants developed a liking, a trust for the character, and a perception that the character cared about their wins and losses and was generally supportive. The conclusion of the study was that "just as people respond to being cared about by other people, users respond to [computer characters] that care."[3]

A robot's social competence, and therefore the way it is perceived by humans as a social being, is inextricably linked to its emotional intelligence.* We saw in chapter 3 that the design of robot dogs benefits from the canine-ethology literature. Similarly, creating an accurate and sophisticated model of human emotion is a task that benefits from the literature on human psychology, and it is unlikely to be many years before all the key elements described in that literature

*Emotional intelligence is defined by Daniel Goleman, the originator of the concept, as "the ability to monitor one's own and others' emotions, to discriminate among them, and to use the information to guide one's thinking and actions."

have been modeled and programmed. Just imagine how powerful these combined technologies will become a few decades from now—speech, vision, emotion, conversation—when each of them has been taken to a humanlike level, a level that today is only a dream for AI researchers. The resulting combination will be an emotional intelligence commensurate with that of a sophisticated human being. The effect will be sensational.

Even though computers have such a wide range of capabilities that they are already pervasive throughout many aspects of our lives, they are not yet our intellectual and emotional equals in every respect, and they are not yet at the point where human-computer friendships can develop in a way that mirrors human-human friendships. Perhaps the strongest influence on the attitudes of those who do *not* believe in a future populated with virtual friends is their difficulty in relating to an artifact, an object that they know is not alive in the sense we usually employ the word. I do not for a moment expect all this to change overnight, and until computer models of emotion and personality are sufficiently advanced to enable the creation of high-quality virtual minds on a par with those of humans, it seems to me inevitable that there will be many who doubt the potential of robots to be our friends. At the present time, we are happy (or at least most of us are) with the idea of robots assembling our cars, robots mowing our lawns and vacuuming our floors, and with robots playing a great game of chess, but not with robots as baby-sitters or robots as intimate friends. Yet the *concept* of robots as baby-sitters is, intellectually, one that ought to appeal to parents more than the idea of having a teenager or similarly inexperienced baby-sitter responsible for the safety of their infants. The fundamental difference at the present time, between this responsibility and that of building cars or playing grandmaster level chess, is surely that robots have not *yet* been shown to be capable baby-sitters, whereas they *have* been shown to excel on the assembly line and on the chessboard. What is needed to convert the unbelievers is simply the proof that robots can indeed take care of the security of our little ones better than we can. And why not? Their smoke-detection capabilities will be better than ours, and they will never be distracted for the brief moment it can take for an infant to do itself some terrible damage or be snatched by a deranged stranger.

One example of how a strong disbelief and lack of acceptance for intelligent computer technologies can change to a diametrically opposite viewpoint has been seen in the airline industry, with automatic pilots on passenger planes. When I was first an airline passenger, around 1955, we had the comfort of seeing the captain of the aircraft walking through the cabin nodding a hello to some of the passengers and stopping to chat with others while his co-pilot took the controls. There was something reassuring about this humanization of the process of flying, to know that people with such obvious authority and the nice uniforms to match were up at the front ensuring that our takeoffs and landings were safe and negotiating the plane securely through whatever storms and around whatever mountain ranges might pose some risk of danger. In those days if all airline passengers had been offered the choice between having an authoritative human pilot in charge and having a computer responsible for their safety, I feel certain that the vast majority would have preferred the human. But today, fifty-plus years later, the situation is very different. Computers have been shown to be so superior to human pilots in many situations that there have been prosecutions brought in the United States against pilots who did *not* engage the computer system to fly their aircraft when they should have done so. This about-face, from a lack of confidence in the capabilities of a computer to an insistence that the computer is superior to humans at the task, will undoubtedly occur in many other domains in which computer use is being planned or already implemented, including the domain of relationships. The time will come when instead of a parent's asking an adolescent child, "Why do you want to date such a schmuck?" or "Wouldn't you feel happier about going to the high school prom with that nice boy next door?" the gist of the conversation could be, "Which robot is taking you to the party tonight?" And as the acceptability of sociable robots becomes pervasive and they are treated as our peers, the question will be rewritten simply as, "Who's taking you to the party tonight?" Whether it is a robot or a human will become almost irrelevant.

Different people will of course adapt to the emotional capacities of robots at different rates, depending largely on a combination of their attitude and their experience with robots. Those who accept that computers (and hence robots) already possess or will come to possess humanlike psychological and mental capabilities will be the first converts. But those who argue that a computer "cannot have emotions" or that robots will "never" have humanlike personalities will probably remain doubters or unbelievers for years, until well after many of their friends have accepted the concept and embraced the robot culture. Between those two camps, there will be those who are open-minded, willing to give robots a try and experience for themselves the feelings of amazement, joy, and emotional satisfaction that robots will bring. I believe that the vast majority in this category will quickly become converts, accepting the concept of robots as relationship partners for humans.

Bill Yeager suggests that this level of acceptance will not happen overnight, because the breadth and depth of the human experience currently go far beyond the virtual pets and robots made possible by the current state of artificial intelligence. As long as robots are different enough from us to be regarded as a novelty, our relationships with them will to some extent be superficial and not even approach the relationships we have with our pets. One of the factors that cause us to develop strong bonds with our (animal) pets is that they share our impermanence, our frailties, being caught up in the same life-death cycle that we are. Yeager believes that to achieve a level of experience comparable with that of humans, robots will have to grow up with us; acquire our experiences with us; be our friends, mates, and companions; and die with us; and that they will be killed in automobile accidents, perhaps suffer from the same diseases, get university degrees, be dumb, average, bright, and geniuses.

I take a different view. I believe that almost all of the experiential benefits that Yeager anticipates robots will need can either be designed and programmed into them or can be compensated for by other attributes that they will possess but we do not. Just as AI technologies have made it possible for a computer to play world-class chess, despite thinking in completely different ways from human grandmasters, so yet-to-be-developed AI technologies will make it possible for robots to *behave* as though they had enjoyed the full depth and breadth of human experience without actually having done any such thing. Some might be skeptical of the false histories that such behavior will imply, but I believe that the behavior will be sufficiently convincing to minimize the level of any such skepticism or to encourage a robot's owner to rationalize its behavior as being perhaps influenced by a previous existence (with the same robot brain and memories but in a different robot body).

I see the resulting differences between robots and humans as being no greater than the cultural differences between peoples from different countries or even from different parts of the same country. Will robots and humans typically interact and empathize with one another any less than, say, Shetland Islanders with Londoners, or the bayou inhabitants of Louisiana with the residents of suburban Boston?

]]]]] Preferring Computers to People

Many people actually *prefer* interacting with computers to interacting with other people. I first learned of this tendency in 1967, in the somewhat restricted domain of medical diagnosis. I was a young artificial-intelligence researcher at Glasgow University, where a small department had recently started up—the Department of Medicine in Relation to Mathematics and Computing. The head of this department, Wilfred Card, explained to me that his work into computer-aided diagnosis took him regularly to the alcoholism clinic at the Western Infirmary, one of Glasgow's teaching hospitals. There he would ask his patients how many alcoholic beverages they usually drank each day, and his computer program would ask the same patients the same question on a different day. The statistics proved that his patients would generally confess to a significantly higher level of imbibing when typing their alcohol intake on a teletype* than when they were talking to the professor. This phenomenon, of people being more honest in their communication with comput-

*An early form of computer keyboard.

ers than they are to humans, has also been found in other situations where questions are asked by a computer, such as in the computerized interviewing of job applicants. Another example stems from a survey of students' usage of drugs, investigated by Lee Sproull and Sara Kiesler at Carnegie Mellon University, in which only 3 percent of the students admitted to using drugs when the survey was conducted with pencil and paper, but when the same survey was carried out by e-mail, the figure rose to 14 percent.

A preference for interacting with a computer program that appeared sociable rather than with a person was observed a year or so after Card's experience by Joseph Weizenbaum at MIT, when a version of his famous ELIZA program was run on a computer in a Massachusetts hospital. ELIZA's conversational skills operated simply by turning around what a user "said" to it, so that if, for example, the user typed, "My father does not like me," the program might reply, "Why does your father not like you?" or "I'm sorry to hear that your father doesn't like you."* Even though ELIZA was dumb, with no memory of the earlier parts of its conversation and with no understanding of what the user was saying to it, half of those who used it at the hospital said that they preferred interacting with ELIZA to interacting with another human being, despite having been told very firmly by the hospital staff that it was only a computer program. This stubbornness might have arisen from the fact that the patients knew they were not being judged in any way, since they would have assumed, correctly in this case, that the program did not have any judgmental capabilities or tendencies.

The preference for interacting with computers rather than with humans helps to explain why computers are having an impact on social activities such as education, guidance counseling, and psychotherapy. As long ago as 1980, it was found that a computer could serve as an effective counselor and that its "clients" generally felt more at ease communicating with the computer than with a human counselor. Sherry Turkle describes this preference as an

*See also the section "On Anthropomorphism," page 74.

infatuation with the challenge of simulated worlds. . . . Like Narcissus and his reflection, people who work with computers can easily fall in love with the worlds they have constructed or with their performances in the worlds created for them by others.[4]

Communicating information is by no means the only task for which people prefer to interact with a computer rather than with another human being. It was also noticed in early studies of human-computer interaction that people are generally as influenced by a statement made by a computer as they are when the same statement is made by a human and that the more someone interacts with a computer the more influential that computer will be in convincing the person that it is telling the truth.

I strongly suspect that the proportion of men preferring interaction with computers to interaction with people is significantly higher than the proportion of women, though I'm not aware of any quantitative psychology research in this area. Evidence from the *McGill Report,* for example, shows men to be more prone than women to eschewing human friendships, leaving men with more time and inclination than women to relate to computers. This bias, assuming that it does exist, suggests that men will always be more likely than women to develop emotional relationships with robots, but although this might be the case in the early years of human-robot emotional relationships, I suspect that in the longer term, women will embrace the idea in steadily increasing numbers. One reason, as will be discussed in chapters 7 and 8, is that women will be extremely enthusiastic about robot sex, once the practice has received good press from the mainstream media in general and women's magazines in particular, and in their robot sexual experiences, women will, more than men, want a measure of emotional closeness with their robot. Another scenario that I foresee as being likely is that from the positive publicity about human-robot relationships women who are in or who have recently left a bad relationship will come to realize that there's more than one way of doing better. Yes, it would be very nice to start a relationship with a new man, but one can never be sure how it's going to work out. I believe that having emotional relationships with robots will come to be perceived as a more dependable way to assuage one's emotional needs, and women will be every bit as enthusiastic as men to try this out. In today's world, there are many women, particularly the upwardly mobile career-minded sort, who would have more use for an undemanding robot that satisfied all of their relationship needs than they would for a man.

What is the explanation for the preference of interacting with a computer over interacting with people? The feeling of privacy and the sense of safety that it brings make people more comfortable when answering a computer and hence more willing to disclose information. And some psychologists explain why people often prefer computers to people and can develop a strong affection for computers by describing this form of affection as an antidote to the difficulties many people face in forming satisfactory human relationships. While this is undoubtedly true in a significant proportion of cases, there are also many people who enjoy being with computers simply because computers are cool, they're fun, they empower us.

]]]]] Robotic Psychology and Behavior

The exploration of human-robot relationships is very much a new field of research. While the creation of robots and the simulation of human-like emotions and behaviors in them are fundamentally technological tasks, the study of relationships between humans and robots is an even newer research discipline, one that belongs within psychology. This field has been given the name "robotic psychology" and practitioners within the field are known as "robopsychologists." Among those who have taken a lead in developing this nascent science are a husband-and-wife team at Georgetown University's psychology department, Alexander and Elena Libin, who are also the founders of the Institute of Robotic Psychology and Robotherapy in Chevy Chase, Maryland.

The Libins define robotic psychology as "a study of compatibility between robots and humans on all levels—from neurological and sensory-motor to social orientation."[5] Their own research into human-robot communication and interaction, although still in its infancy, has already demonstrated some interesting results. They conducted experiments to investigate people's interactions with NeCoRo, a sophisticated robotic cat covered with artificial fur, manufactured by the Omron Corporation and launched in 2001. NeCoRo stretches its body and paws, moves its tail, meows, and acts in various other catlike ways, getting angry if someone is violent to it and expressing happiness when stroked, cradled, and treated with lots of love. Additionally, NeCoRo's software incorporates learning methods that cause the cat to become attracted to its owner day by day and to adjust its personality to that of its owner. One of the Libins' earliest experiments was designed to investigate how biological factors such as age and sex, psychological factors such as a person's past experiences with real pets and with technology, and cultural factors such as the social traditions that affect people's communication styles influence the way a person interacts with such a robot.

This experiment found that older people get more pleasure from the responses of the robot cat (its "meows") than do younger people when they touch it. This was attributed to the fact that younger people use cell phones, computers, and household devices more intensively than their elders do and generally experience a greater enjoyment of technology. Another finding was that men get more pleasure than do women from playing with NeCoRo, generally experiencing more excitement when the cat turns its head, opens and closes its eyes, and changes its posture. This bias seems likely to be a symptom of the fact that men, more than women, enjoy interaction with computers, though further research is necessary to test this assumption. Similarly, further experiments will be needed to explain another of the Libins' results: that the American subjects in their experiment enjoyed touching the cat more and obtained more pleasure from the way the cat cuddled them when they were stroking it than did the Japanese subjects. This could be because cats are more popular as pets in American homes than they are in Japan, an explanation given credence by yet another of the Libins' experimental findings, that the degree to which someone likes pets influences the way that they interact with the robotic cat and the enjoyment received from picking it up and stroking it.

Experimental results such as these will help guide robopsychologists toward a greater understanding of human-computer and human-robot interactions, by providing data to assist the robot designers of the future in their goal of making robots increasingly acceptable as friends and partners for humans. As the human and artificial worlds continue to merge, it will become ever more important to study and understand the psychology of human-robot interaction. The birth of this new area of study is a natural consequence of the development of robot science. Our daily lives bring us more frequent interaction with different kinds of robots, whether they be Tamagotchis, robot lawn mowers, or soccer-playing androids. These robots are being designed to satisfy different human needs, to help in tasks such as education and therapy, tasks hitherto reserved for humans. It is therefore important to study the behavior of robots from a psychological perspective, in order to help robot scientists improve the interactions of their virtual creatures with humans.

Much of the early research in this field has been carried out with children, as this age group is more immediately attracted to robot pets than are their parents and grandparents. One of the first findings from this research was intuitively somewhat obvious but nevertheless interesting and useful in furthering good relations between robots and humans. It was discovered that children in the three-to-five age group are more motivated to learn from a robot that moves and has a smiling face than from a machine that neither moves nor smiles. As a result of recognizing these preferences, the American toy giant Hasbro launched a realistic-looking animatronic robot doll called My Real Baby that had soft, flexible skin and other humanlike features. It could exhibit fifteen humanlike emotions by changing its facial expressions—moving its lips, cheeks, and forehead—blinking, sucking its thumb, and so forth. By virtue of these features, it could frown, smile, laugh, and cry.

The appeal to children of My Real Baby lies in its compatibility with them, a compatibility that breeds companionship. And the shape and appearance of a robot can have a significant effect on the level of this compatibility. A study at the Sakamoto Laboratory at Ochanomizu University in Japan investigated people's perceptions of different robots—the AIBO robotic dog and the humanoid robots ASIMO and PaPeRo—and explored how these perceptions compared with the way the same group of people perceive humans, animals, and inanimate objects. One conclusion of the study was that appearance and shape most definitely matter—people feel more comfortable when in the company of a friendly-shaped, humanlike robot than when they are with a robotic dog.

In chapter 3 we discussed the use of ethology, the study of animals in their natural setting, as a basis for the design and programming of robot animals. Since humans are also a species of animal, it would seem logical to base the design and programming of humanoid robots on the ethology of the human species, but unfortunately the ethological literature for humans is nowhere near as rich as it is for dogs, and what literature there is on human ethology is mainly devoted to child behavior. For this reason the developers of Sony's SDR humanoid robot have adapted the ethological architecture used in the design of AIBO, an architecture that contains components for perception, memory, and the generation of animal-like behavior patterns, adding to it a thinking module* to govern its behavior. SDR also incorporates a face-recognition system that enables the robot to identify the face of a particular user from all the faces it has encountered, a large-vocabulary speech recognition system that allows it to recognize what words are being spoken to it, and a text-to-speech† synthesizer allowing it to converse using humanlike speech.

]]]]] Emotions in Humans and in Robots

Building a robot sufficiently convincing to be almost completely indistinguishable from a human being—a Stepford wife, but without her level of built-in subservience—is a formidable task that will require a combination of advanced engineering, computing, and artificial-intelligence

---

*Referred to by its designers as a "deliberative layer."
†Text-to-speech is a speech-synthesis technology that allows the software to say any word, based on its spelling and its assumed pronunciation. It is not therefore limited only to a fixed, preprogrammed vocabulary.

skills. Such robots must not only look human, feel human, talk like humans, and react like humans, they must also be able to think, or at least to simulate thinking, at a human level. They should have and should be able to express their own (artificial) emotions, moods, and personalities, and they should recognize and understand the social cues that we exhibit, thereby enabling them to measure the strengths of our emotions, to detect our moods, and to appreciate our personalities. They should be able to make meaningful eye contact with us and to understand the significance of our body language. From the perspective of engendering satisfying social interaction with humans, a robot's social skills—the use of its emotional intelligence—will probably be even more important than its being physically convincing as a replica human.

Lest I be accused of glossing over a fundamental objection that some people have to the very idea that machines can have emotions, I shall here summarize what I consider to be the most important argument supporting this notion.* Certainly there are scholars whose views on this subject create doubts in the minds of many: How can a machine have feelings? If a machine does not have feelings, what value can we place on its expressions of emotion? What is the effect on people when machines "pretend" to empathize with their emotions? All of these doubts and several others have attracted the interest of philosophers for more than half a century, helping to create something of a climate of skepticism.

To my mind all such doubts can be assuaged by applying a complementary approach like that of Alan Turing when he investigated the question, "Can machines think?"† Turing is famous in the history of computing for contributions ranging from leading the British team that cracked the German codes during World War II to coming up with the

---

*Philosophers have been debating various arguments on this topic since the 1950s at least. One prominent philosopher, Sidney Hook, observed in 1959 that when robots claim they have feelings, our acceptance of their claims will depend on "whether they look like and behave like other people we know." This argument is very similar to the one presented here.

†In 1950, Turing asked this question in his famous paper "Computing Machinery and Intelligence," arguably the most important publication in the history of artificial intelligence.

solution to a number of fundamental issues on computability. But it was his exposition of what has become known as the "Turing test" that has made such a big impact on artificial intelligence and which enables us, in my view, to answer all the skeptics who pose questions such as, "Do machines have feelings?"

The Turing test was proposed as a method of determining whether a machine should be regarded as intelligent. The test requires a human interrogator to conduct typed conversations with two entities and then decide which of the two is human and which is a computer program. If the interrogator is unable to identify the computer program correctly, the program should be regarded as intelligent. The logical argument behind Turing's test is easy to follow—conversation requires intelligence; ergo, if a program can converse as well as a human being, that program should be regarded as intelligent.

To summarize Turing's position, if a machine gives the *appearance* of being intelligent, we should assume that it is indeed intelligent. I submit that the same argument can equally be applied to other aspects of being human: to emotions, to personality, to moods, and to behavior. If a robot behaves in a way that we would consider uncouth in a human, then by Turing's standard we should describe that robot's behavior as uncouth. If a robot acts as though it has an extroverted personality, then with Turing we should describe it as being an extrovert. And if, like a Tamagotchi, a robot "cries" for attention, then the robot is expressing its own form of emotion in the same way as a baby does when it cries for its mother. The robot that gives the *appearance,* by its behavior, of having emotions should be regarded as *having* emotions, the corollary of this being that if we want a robot to appear to have emotions, it is sufficient for it to *behave* as though it does. Of course, a robot's programmed emotions might differ in some ways from human emotions, and robots might even evolve their own emotions, ones that are very different from our own. In such cases, instead of understanding, through empathy and experience, the relationship of a human emotion to the underlying causes, we might understand nothing about robotic emotions except that on the surface they resemble our own. Some people will not be able to empathize with a robot that is frowning or grinning—they will be people who interpret the robot's behavior as nothing more than an act, a performance. But as we come to recognize the various virtual emotions and experiences that lie behind a robot's behavior, we will feel less and less that a robot's emotions are artificial.

Our emotions are inextricably entwined with everything we say and do, and they are therefore at the very core of human behavior. For robots to interact with us in ways that we appreciate, they, too, must be endowed with emotions, or at the very least they must be made to *behave* as though they have emotions. Sherry Turkle has found that children deem simple toys, such as Furby, to be alive if they believe that the toy loves them and if they love the toy. On this basis the perception of life in a humanoid robot is likely to depend partly on the emotional attitude of the user. If users believe that their robot loves them, and that they in turn love their robot, the robot is more likely to be seen as alive. And if a robot is deemed to be alive, it is more likely that its owner will develop increased feelings of love for the robot, thereby creating an emotional snowball. But before robot designers can mimic emotional intelligence in their creations, they must first understand human emotions.

Human emotions are exhibited in various ways—in the changes in our voice, in the changes to our skin color when we blush, in the way we make or break eye contact—and robots therefore need similar cues to help express *their* emotions. Just as face and sound are used as a matter of course, instinctively and subconsciously, by humans communicating with other humans, so similar forms of communication are being exhibited by emotionally expressive robots to communicate their simulated emotions to their human users.

Many studies have shown that the activity of the facial muscles in humans is related to our emotional responses. The muscle that draws up the corners of the lips when we smile* is associated with positive experiences, while the muscle that knits and lowers the brows when we frown† is associated with negative ones. Much of today's research

---

*This muscle is called zygomaticus major.
†This muscle is corrugator supercilii.

into the use of facial expression in computer images and robots stems from a coding system developed during the 1970s by Paul Ekman, a psychologist at the University of California at San Francisco. Ekman classified dozens of movements of the facial muscles into forty-four "action units"—components of emotional expression—each combination of these action units corresponding to a different variation on a basic facial expression such as anger, fear, joy, or surprise. It has been shown as a result of Ekman's work that the creation of emotive facial expressions is relatively easy to simulate in an animated character or a robot, while research at MIT has revealed that humans are capable of distinguishing even simple emotions in an animated character by observing the character's facial expressions. The recognition, by a machine, of these various action units can therefore be converted to the recognition of a human emotional state. And the simulation of a combination of action units becomes the simulation, in a robot or on a computer screen, of a human emotion. Yes, this is an act on the part of the robot, but as time goes on, the act will become increasingly convincing, until it is so good that we cannot tell the difference.

The study of emotions and other psychological processes is a field that predates the electronic computer, providing researchers in robotics with a pool of research into which they can tap for ideas on how best to simulate these processes in robots. If we understand how a particular psychological process works in humans, we will be able to design robots that can exhibit that same process. And just as being human endows us with the potential to form companionable relationships, this same potential will be designed into robots to help make them sociable. Some would argue that robot emotions cannot be "real" because they have been designed and programmed into the robots. But is this very different from how emotions work in people? We have hormones, we have neurons, and we are "wired" in a way that creates our emotions. Robots will merely be wired differently, with electronics and software replacing hormones and neurons. But the results will be very similar, if not indistinguishable.

An example of a robot in which theories from human psychology have been synthesized is Feelix, a seventy-centimeter-tall humanoid robot designed at the University of Århus and built with Lego bricks. The manner in which a user interacts with Feelix is by touching its feet. One or two short presses on the feet make Feelix surprised if they immediately follow a period of inactivity, but when the presses become more intense and shorter, Feelix becomes afraid, whereas a moderate level of stimulation, achieved by gentle, long presses on its feet makes Feelix happy. But if the long presses become more intense and sustained, Feelix becomes angry, reverting to a happier state and a sense of relief only when the anger-making stimulation ceases.

Feelix was endowed with five of the six "basic emotions" identified by Paul Ekman: anger, fear, happiness, sadness, and surprise.* All five emotions have the advantage that they are associated with distinct corresponding facial expressions that are universally recognized, making it possible to exhibit the robot's emotions partly by simulating those facial expressions. Anger, for example, is exhibited by having Feelix raise its eyebrows and moderately opening its mouth with its upper lip curved downward and its lower lip straight, while happiness is shown by straight eyebrows and a wide closed mouth with the lips bent upward. When it feels no emotion—that is, when none of its emotions are above their threshold level, Feelix displays a neutral face. But when it is stimulated in various ways, Feelix becomes emotional and displays the appropriate facial expression.

In order to determine how well humans can recognize emotional expressions in a robot's face, Feelix was tested on two groups of participants, one made up of children in the nine-to-ten age range and one with adults aged twenty-four to fifty-seven. The tests revealed that the adults correctly recognized Feelix's emotion from its facial expression in 71 percent of the tests, with the children slightly less successful at 66-percent recognition. These results match quite well the recognition levels demonstrated in earlier tests, using photographs of facial expressions, that had been reported in the literature on emotion recognition, providing evidence that the simulation of expression of the basic emo-

---

*The sixth emotion proposed by Ekman, disgust, was not felt appropriate for the type of interactions that Feelix's designers expected humans to have with the robot.

tions is not something from science fiction but can already be designed into robots. Accepting that an acted-out emotion is just that, an act, will make it difficult to believe that the acted emotion is being experienced by the robot. But again, as the "acting" improves, so any disbelief will evaporate.

]]]]] Robot Recognition of Human Emotions

To interact meaningfully with humans, social robots must be able to perceive the world as humans do, sensing and interpreting the same phenomena that humans observe. This means that in addition to the perception required for physical functions such as knowing where they are and avoiding obstacles, social robots must also possess relationship-oriented perceptual abilities similar to those of humans, perception that is optimized specifically for interacting with humans and on a human level. These perceptual abilities include being able to recognize and track bodies, hands, and other human features; being capable of interpreting human speech; and having the capacity to recognize facial expressions, gestures, and other forms of human activity.

Even more important than its physical appearance and other physical attributes in engendering emotional satisfaction in humans will be a robot's social skills. Possibly the most essential capability in robots for developing and sustaining a satisfactory relationship with a human is the recognition of human emotional cues and moods. This capability must therefore be programmed into any robot that is intended to be empathetic. People are able to communicate effectively about their emotions by putting on a variety of facial expressions to reflect their emotional reactions and by changing their voice characteristics to express surprise, anger, and love, so an empathetic robot must be able to recognize these emotional cues.

Robots who possess the capability of recognizing and understanding human emotion will be popular with their users. This is partly because, in addition to the natural human desire for happiness, a user might have other emotional needs: the need to feel capable and competent, to maintain control, to learn, to be entertained, to feel comfortable and supported. A robot should therefore be able to recognize and measure the strength of its user's emotional state in order to understand a user's needs and recognize when they are being satisfied and when they are not.

Communicating our emotions is a process called "affect," or "affective communication," a subject that has been well investigated by psychologists. It is also a subject of great importance in the design of computer systems and robots that detect and even measure the strength of human emotions and in systems that can communicate their own virtual emotions to humans. The Media Lab at MIT has been investigating effective communication since the mid-1990s, in research led by Rosalind Picard, whose book *Affective Computing* has become a classic in this field. Affective computing involves giving robots the ability to recognize our emotional expressions (and the emotional expressions of other robots), to measure various physiological characteristics in the human body, and from these measurements to know how we are feeling.

Inexpensive and effective technologies that enable computers to measure the physiological indicators of emotion also allow them to make judgments about a user's emotional state. Thanks largely to Picard, detecting and measuring human emotion has become a hot research topic in recent years. By measuring certain components of the human autonomic nervous system,* it is already possible for computers to distinguish a few basic emotions. A simple example of such measurements is galvanic skin response—the electrical conductivity of the skin. This has long been known as an indicator of stress and has therefore been employed in some lie detectors, but more recently it has also been used as a metric for helping to recognize certain emotional states other than stress. Heart rate is another easy-to-measure example—it is known to increase most during fear but less when a person is experiencing anger, sadness, happiness, surprise, and disgust,

---

*The autonomic nervous system is that part of the vertebrate nervous system that regulates involuntary action—for example, the actions of the intestines, the heart, and the glands.

the last of these eliciting only the barest minimum of a heart-rate change. Yet another example is blood pressure, which increases during stress and decreases during relaxation, the biggest increase again being associated with anger.

It is a relatively simple matter to measure human blood pressure, respiration, temperature, heart rate, skin conductivity, and muscle tension using what are currently regarded as sophisticated items of electronic equipment. Research into "affective wearables," usually items of clothing and other attachments that may be worn unobtrusively and come with electronic sensors for taking such measurements, will inevitably lead to the development of technologies that can monitor all these vital signs without our even noticing that we're wearing them. By transmitting the measured data, affective wearables will thus enable robots to recognize and quantify at least some of our emotions, allowing them to judge our moods, based on our displays of emotion as they appear to the electronic monitors. For example, by combining the data from only four different measures—respiration, blood pressure volume, skin conductance, and facial-muscle tension—Rosalind Picard, Elias Vyzas, and Jennifer Healey developed an emotion-recognition system capable of 81-percent accuracy when distinguishing among eight emotions: anger, hate, grief, platonic love, romantic love, joy, reverence, and the neutral state (no emotion).

Additional help in detecting human emotion can come from auditory and visual cues. Facial-recognition technology is making dramatic advances, spurred on by the impetus of a fear of terrorism—the technology that today successfully identifies faces seen on a closed-circuit TV camera will tomorrow be identifying not only the person behind the face but also that person's mood. Similarly with voices. Voice recognition has taken on increased import as a means of identification for security purposes, turning the sound characteristics of the human voice into measurable quantities that can act as an additional aid to identification. Iain Murray and John Arnott have investigated the vocal effects associated with several basic emotions, establishing links between voice characteristics and emotion that make possible the design of a voice-based emotion recognizer. This particular slant on the technology comes from the measurement of the pitch of a voice, the speed with which words are uttered, the frequency range of the voice, and changes in volume. Someone who is sad or bored will typically exhibit slower, lower-pitched speech, while a person who is afraid, angry, or joyous will speak louder and faster, with more words spoken at higher frequencies.

In summary, the creation of natural and efficient communication between human and robot requires that each display emotions in ways that the other is able to recognize and assess. But the emotionally intelligent robot must not only be able to recognize emotions in humans and to assess the strength of those emotions, it should also *demonstrate* that it recognizes the emotions displayed by its human. As the development of emotion-recognition and emotion-simulation technologies advances, so will the development of emotional intelligence in robots, and their relationships with humans will come to mirror a healthy human-human relationship.

]]]]] Three Routes to Falling in Love with Robots

There are three distinct progressions that I believe will lead enormous numbers of humans to develop affection for and fall in love with robots. One route will develop in a humanlike loving way, as robots become more and more human in appearance and personality, encouraging us to like and to love them. This is a natural extension of normal human loving and is the easiest of the three routes to comprehend. And, just as with the Tamagotchi, the human tendency to nurture will help to engender in us feelings of love for robots.

Another route is via a love for machines and technology per se, sometimes called "technophilia." People who "love" computers and machines do so in different ways. There are those who rush out and buy every new technological gizmo the moment it is put on sale—theirs is a love for all new technology. There are those for whom the technology converts into some other form of emotional or even erotic stimulation, such as pornography on the Internet or on a DVD. There are the technophiles, usually programmers but also those who love pressing buttons to make their gizmos do weird and wonderful things; theirs is a love of control, whether it is control by writing the programs that instruct their computers what to do or the much simpler form of control achieved by pressing the buttons on devices that have already been programmed. And the act of programming has itself been compared to sex, in that programming is a form of control, of bending the computer or the gadget to the will of the programmer, forcing the computer to behave as one wishes—domination.

A love of technology and its benefits was at first largely the province of the technically more adept, the economically upward mobile, and, predominantly, of adolescents and those in their twenties and thirties. As the cost of electronics has come down, enabling consumer-electronics manufacturers to create electronic toys and other products especially for children, so the age range of technophiles has widened considerably. Nowadays, with primary-school children and even preschoolers finding themselves the owners of a plethora of electronic products, we are creating future generations of adults for most of whom the latest gizmos will seem perfectly normal rather than amazing. And so it will be with robotics. Those who are born surrounded by electronics will grow up eager for and receptive to whatever new electronic inventions become available during their lifetimes. The love that yesterday's children and young adults demonstrated for their Furbies and Tamagotchis will be the basis for the adults of the future to find it perfectly normal first to love their interactions with robots and then to love the robots themselves.

The evolution of loving relationships between humans and robots will be yet another example of how technology changes the way we live in dramatic, even mind-boggling ways. One of the most glaring examples from the twentieth century is television. Who at the time of the First World War would imagine that one day they would be able to look at a box that showed something happening, at that very moment, on the other side of the world, or even on the moon? Who at the time of the

Second World War would have believed that by the end of the century telephone booths in the street would fast become redundant, because just about everyone would be walking around with their very own wireless telephone in their pocket? Who at the time of the Vietnam War would have expected handwritten letters to gradually go out of fashion in the United States and many other countries as more and more people would take to the computer as their primary or sole means of writing letters and even sending them, at virtually no cost, to their friends and relatives, in no more time than it takes to click a computer mouse? And which did you, dear reader, use more recently as a source of information, a reference book or an Internet search engine such as Google?

The entertainment industry has been reshaped more than most by the tools of technology. Animation, made so popular for generations of children (and adults) by Walt Disney and originally hand-drawn, painstakingly, by teams of artists, is nowadays created automatically by superfast computers, costing animators their jobs by the thousands. Music that in my youth came into our homes on gramophone records that rotated at 78 revolutions per minute, and later at 45 and then 33 rpm, the slower speeds allowing more music to be stored on a single disc, now comes to our handheld boxes by "download" via the Internet, making available to us a colossal collection of pop, rock, jazz, classical, and all other types of music without our having any need to go to a store. And then there are video games, probably the biggest-ever product success in the entertainment industry—games that today offer the user the most amazing sights, sounds, and action, all in an easily portable package. Other video-based products such as DVDs and their precursors—videocassettes—have also created huge changes in the way we entertain ourselves, enabling us to have the movies of our choice in our homes, to watch and watch again as often as we wish. (And the genre that has achieved the biggest financial success in that particular technological field is pornographic movies, because sex seems always to find a way to reach the marketplace. Sex sells.)

But back to robots. A third route in the evolution of love for robots will arise out of emotions similar to those that have made Internet relationships so hugely popular. Let us recall Deb Levine's words, quoted in chapter 1:

> For some people, online attraction and relationships will become a valid substitute for more traditional relationships. Those who are housebound or rurally isolated and those who are ostracized from society for any number of different reasons may turn to online relationships as their sole source of companionship.[6]

The same could equally be said of human-robot relationships, and some will find this worrying. Most people who develop emotional attachments to robots, and to whom robots exhibit their own demonstrations of love, will have in their mind the knowledge that the robot is just that, a robot, and not a human being. This "you are only a robot" syndrome will be some kind of boundary across which a human must pass to feel love to its fullest extent for a robot, though in the case of certain groups within our society, crossing that boundary will seem perfectly natural. Those who prefer to relate to computers rather than to humans will doubtless find it no problem at all. Nor will many nerds, many social outcasts, and those who will be only too happy to find someone, almost anyone, who exhibits affection for them. But what about the more normal members of the population? What will it take for them to cross this boundary? One could argue that the first requirement will be incredibly good engineering, so that robots are as convincing in their appearance and actions as Stepford wives—almost indistinguishable from humans. But as we saw in chapter 3, the Tamagotchi experience and the reactions of the owners of AIBO pet dogs indicate that very strong emotional attachments *can* develop in humans even when the object of such affection is not humanlike in appearance.

This aside into the world of Internet romances and its implications has another important point to make in my line of argument on the subject of love with robots. One conclusion that can safely be drawn from the phenomenon of falling in love via the Internet, as with a pen pal, is that it is *not* a prerequisite for falling in love ever to be in the presence of the object of one's love. The falling-in-love process can be conducted completely in the physical absence of the loved one. This is consistent with and a much stronger form of the phenomenon noted by Robert Zajonc.* Of course there are photographs and video images of the loved one that can be received via the Internet. And the loved one's voice can be heard via the Internet or the telephone, but their physical presence is simply not necessary.

Now consider the following situation: At the other end of an Internet chat line, complete with a webcam to transmit its image, a microphone to carry the sound of its voice, and a smell-detection and -transmission system to convey its artificial bodily scent to you, there is a humanlike robot endowed with all of the artificially intelligent characteristics that will be known to researchers by the middle of this century. You sit at home looking at this robot, talking to it, and savoring its fragrance. Its looks, its voice, and its personality appeal to you, and you find its conversation simulating, entertaining, and loving. Might you fall in love with this robot? Of course you might. Why shouldn't you? We have already established that people can fall in love *without* being able to see or hear the object of their love, so, clearly, being able to see it, finding its looks to your liking, being able to hear its sexy voice, and being physically attracted by its simulated body fragrance can only strengthen the love you might develop in the absence of sight, sound, and smell.

And if you do fall in love with a robot, what will be the nature of this love and how will it differ from the way you feel about the love of your life in the world as it is today?

As noted earlier, one important difference will be that robots are going to be replicable, even to the point of their personality, their memories, and their emotions. Those readers who are frequent computer users will know that it is good practice to back up your work on the computer just in case of a disaster that causes the loss of some or all of your data. Similarly, it will become common practice for the knowledge, personality, and emotion parameters—and all the other software aspects of a robot's "brain"—to be backed up on a frequent basis. By midcentury this process will almost certainly be fully automatic, so

*See page 32.

that neither the robot nor its owner needs to do anything. At regular intervals the contents of the robot's brain, its consciousness, its emotions, will all be transmitted to a secure memory bank. If, heaven forbid, a robot is damaged or destroyed and its owner wishes an exact copy, the physical characteristics can be replicated in the robot factory, and then the contents of the brain, predamage, can be downloaded into the new copy of the original robot. This capability creates one enormous difference between the love one feels for another human being and the love that will be felt for robots. If you love someone enough, you will willingly undertake any risk, or knowingly sacrifice your own life, in order to save theirs. This is only partly because of the strength of your love for them. It is also partly because they are irreplaceable. But in the case of love for a robot, it will be as though death simply does not exist as a concept that can be applied to the object of your love. And if it can never truly die, because it can always be brought back to life in an exact replica of its original body, there will never be any need for a human to sacrifice their own life for their robot or to take a major risk on its behalf.

Another important difference is that robots will be programmable never to fall out of love with their human, and they will be able to reduce the likelihood of their human falling out of love with them. Just as with the central heating thermostat that constantly monitors the temperature of your home, making it warmer or cooler as required, so your robot's emotion-detection system will constantly monitor the level of your affection for it, and as that level drops, your robot will experiment with changes in behavior aimed at restoring its appeal to you to normal.

]]]]] Robot Personalities and Their Influence on Relationships

Personality is one of the most important factors that drive the processes of falling in love and falling in lust, so before we examine the specific causes of falling in love with robots, we shall first consider some of the significant research on robot personality that has been conducted during the past decade or so.

Robot personality is a subject that some readers might regard with skepticism—how can a robot have a personality? In the mid-1990s, Clifford Nass and some of his colleagues in the Department of Communication at Stanford University showed it to be relatively straightforward to create humanlike characteristics in computers—computer personalities—using a set of cues drawn from the extensive literature on the subject of human personality. In psychological terms, personality is the set of distinctive qualities that distinguish individuals. Nass and his group have conducted more than thirty-five experiments to investigate some of these qualities, to determine how they can be simulated in computer programs and how such simulations compare with the corresponding trait in humans.

One of the experiments carried out by Nass's group is related to the team element of a partnership relationship. Couples act as a team in myriad ways: She might wash the dishes while he dries, she might do the laundry while he does the gardening, he might be the principal breadwinner while she devotes more time to taking care of the children—or vice versa. It is not only the drudge tasks that are shared in a partnership relationship, it is also the more pleasant ones, and in both cases the sharing of responsibilities will often act as a bonding factor, helping to sustain the relationship. A study of computers as teammates is therefore of considerable interest in estimating how a computer-human dyad might also function as a team.

Nass and Byron Reeves based their study into computers as teammates on social-psychology experiments showing that there are two key factors in a team relationship—group identity and group interdependence. Group identity simply means that a team must have something to identify it by, often just a name such as "Mr. and Mrs. Bloggs," or "the Smith family," or "Christine and David." The importance of group interdependence lies in the fact that the behavior of each member of a team can affect all the other members.*

The teams created for this study each consisted of a human and a

---

*In the case of a relationship dyad, the word "all" relates, of course, to both partners in the relationship.

computer, with the team identified by a color and the members of the team sporting a ribbon of that color and a notice saying "blue team," for instance, on that team's computer. Half of the people in the experiment were told they were on the blue team. They were also told that their performance would be graded and that its final evaluation would depend not only on their own efforts but also on those of the blue team's computer. The other half of the people in the experiment were treated as though they were not on the same team as the computer with which they were collaborating. These subjects also wore a blue ribbon, but their computer was dressed in green and carried a notice affirming that it was a "green computer." The experimenters made no mention to the humans in the second group of any collaboration between them and the computer, in order to avoid creating an association of teamwork in their minds. These subjects were told that their performance would be graded solely on the basis of their own work with the computer—that the computer was simply there to help.

The participants were set to work on a problem-solving task commonly employed in experimental psychology, a task known as the Desert Survival Problem.* When the participants first attacked the problem, they would try to solve it by themselves, creating their own ranking for the survival items. They then went into another room, one at a time, where they worked on the task in collaboration with their assigned computer. They all exchanged information with their computer about each of the twelve survival items, and, if they wished, the participants could then change their initial rankings. Once the human participants had interacted with their computer, they would be sent into a third room, where they wrote out their final rankings and responded to questions about their interaction with their computer, questions such as "How similar was the computer's approach to your own approach in evaluating the twelve items?" and "How helpful were the computer's suggestions?"

*This task requires the participants to imagine themselves as copilots of a plane that has crash-landed in the desert and to decide on the order of importance of twelve objects that might help in their survival, such as a quart of water and a flashlight. Each participant in a pair (in this case one computer and one human) exchanges their initial rankings with the partner and discusses each object. These discussions enable experimental psychologists to measure the assertiveness of each participant.

The results of this experiment revealed a lot about how people perceive team relationships. When the humans believed that they were on the same team as the computer, they assessed the computer as being more like themselves relative to how much like themselves the participants thought the computers to be when the participants worked alone. These "teamed" participants also thought that their "teammate" computer had adopted a problem-solving style more similar to their own and that their computer agreed more completely with their own ranking of the items. Another tendency was for the teamed participants to believe that the information given to them by the computer was more relevant and helpful and that it was presented in a friendlier manner compared to the participants who did not believe they were members of a human-computer team—all this despite the fact that the information was identical and was presented in an identical manner in both cases. Other indications of relationship building between the human participants and their computers were that the teamed participants tried harder to reach an agreement with their computer on the rankings and were more receptive to their teammate's suggestions and influences.

One of the most important conclusions of this study was to confirm the work of earlier psychologists who "have long been excited by how little it takes to make people feel part of a team, and by how much is gained when they do." Reeves and Nass had extended this earlier research by showing that feelings of being part of a team are powerful enough to affect people's interactions with computers, once they believe that their own success depends also on the success of the computer.

This research was groundbreaking work at that time, but even more remarkable than the ease with which their goal was accomplished was what the experimenters learned when they tested two simple computer personalities, each designed into a program that collaborated with a human user on the Desert Survival Problem. One of these computer personalities was "dominant," using strong language in its assertions and commands, displaying a high level of confidence when communicating with the human test subjects, and leading off the dialogues with its human collaborators. The other computer personality was "submissive," using weaker language, in which assertions were replaced by suggestions and commands with questions, and inviting or allowing the human collaborator to start each dialogue. It was found that those humans who themselves had more dominant personalities* enjoyed interacting with the dominant computer more than they did with the submissive one, while those with a more submissive personality preferred interacting with the submissive computer. Furthermore, not only did the human subjects *prefer* to interact with a computer similar in personality to their own, but they also experienced a greater satisfaction in their *own* performance on the problem-solving task when collaborating with the similar computer. These results led to the conclusion that not only do humans prefer to interact with other humans of similar personality, but they also prefer to interact with *computers* that have similar (virtual) personalities to their own.

Other experiments conducted by Nass and his group confirmed that humanlike behavior by a computer enhances the user's experience of the interaction and makes the computer more likable. One example of this phenomenon is the ability of computers to increase users' liking of them by means of flattery, by matching the users in personality, and through the use of humor, which has been found to lead to assessments of them as being more likable, competent, and cooperative than computers that do not exhibit any humor. Another example came from highly expressive teaching programs that were found to increase students' feelings of trust in the programs because the students perceived them as helpful, believable, and concerned.

]]]]] Designing Robot Personalities

Designing a robot with an appealing personality is an obvious goal, one that would allow you to go into the robot shop and choose from a range of personalities, just as you will be able to choose from a range of heights,

---

*The personality of each of the human subjects was tested for dominance and submissiveness using a standard personality test commonly employed by psychologists.

looks, and other physical characteristics. One interesting question is whether it will be necessary to program robots to exhibit some sort of personality friction for us to feel satisfied by our relationships with them and to feel that those relationships are genuine. Certainly it would be a very boring relationship indeed in which the robot always performed in exactly the manner expected of it by its relationship partner, forever agreeing with everything that was said to it, always carrying out its human's wishes to the letter and in precisely the desired manner. A Stepford wife. Perfection. No, that would not be perfection, because, paradoxically, a "perfect" relationship requires some imperfections of each partner to create occasional surprises. Surprises add a spark to a relationship, and it might therefore prove necessary to program robots with a varying level of imperfection in order to maximize their owner's relationship satisfaction. Many people have relatively stable personalities and would therefore probably appreciate robots whose own personality and behavior exhibited some, but not a huge amount of, perturbation. This variable factor in the stability of a robot's personality and emotional makeup is yet another of the characteristics that can be specified when ordering a robot and that can be modified by its owner after purchase. So whether it is mild friction that you prefer or blazing arguments on a regular basis, your robot's "friction" parameter can be adjusted according to your wishes. Your robot will be programmed to recognize and measure friction when it is there, by the nature of your conversation with it and the tone of your voice, and to increase or decrease the level of friction according to your preferences.

One important consideration for robot programmers when planning a robot's personality and behavior will be how best to cope with different cultures. Just think of the courting rituals and the chaperone phenomena in some Latin countries, the Chinese tendency not to be too physically demonstrative in public and the contrasting lack of inhibitions displayed in some other countries, and the tradition of arranged marriages in certain cultures—a tradition that ought to present no problem for robots, because the parents of the human bride or groom will simply make all the choices in the robot shop as to its physical appearance and other characteristics, rather than leave these decisions to their offspring. Whatever the social norms of the prospective owners and their culture, a robot will be able to satisfy them. Similarly with religion, the details and intensity of which can be chosen and changed at will—whether you're looking for an atheist, an occasional church-goer, or a devout member of any religion, you have only to specify your wishes when placing your order at the robot shop. The key here will be ensuring that the robot has a flexible personality. It will most likely leave the factory with a set of personality traits, some standard and others chosen by the customer, but a robot will be able to set any or all of these traits aside as required, to allow the robot itself to adapt to the personality needs of its owner.

The example of the dominant and submissive problem-solving programs devised by Nass and his team suggests that creating artificial personalities will probably not be an immensely difficult task for robot scientists. Likewise, the creation of blue eyes, a sexy voice, or whatever other physical characteristics turn you on, are all within the bounds of today's technology. And if what turned you on when you purchased your robot ten years ago no longer turns you on today, the adaptability of your robot and the capability of changing any of its essential charac-teristics will ensure that it retains your interest and devotion. When robots are able to exhibit the whole gamut of human personality and physical characteristics, their emotional appeal to humans will have reached a critical level in terms of attracting us, inducing us to fall in love with them, seducing us in the widest sense of the word. We will recognize in these robots the same personality characteristics we notice when we are in the process of falling in love with a human. If someone finds a sexy voice in their partner a real turn-on, they are likely to do so if a similar voice is programmed into a robot. If it's blue eyes that one is after, simply select a blue-eyed robot when you make your choice. If it's a particular personality trait, your robot will come with that trait ready-made, or it will learn the trait as it discovers its importance to you.

While much of the development work on the hardware for new robot technologies is being carried out in Japan, the West is not lagging behind in the research effort into software for the robots' emotions and personality.* One reason for the Japanese bias toward hardware is because the Japanese government is determined to employ robots in the future to assist with the massive task of taking care of their aging population, a task for which the hardware must be totally reliable and robust. Another motivation for the Japanese investment in robot hardware research is that it will be the Japanese consumer-electronics conglomerates that will reap the greatest commercial benefits when robots are on sale to the public in high-volume quantities.

These world leaders in robotics, Japan and the United States, have somewhat different approaches and goals. The United States produces and uses far fewer robots than does Japan, because the United States is more reliant on less expensive immigrant labor. According to the latest industry figures in 2006, the United States had only 68 robots in manufacturing industries for every 10,000 human manufacturing workers, whereas Japan had 329 per 10,000. But an even greater distinction lies in the cultural differences between Japan and the United States and how these differences transfer to the different perceptions of the people in these countries to the prospect of our future with robots.

In an article in *USA Today*,[†] Kevin Maney summarizes these differences:

U.S. labs and companies generally approach robots as tools. The Japanese approach them as beings. That explains a lot about robot projects coming out of Japan.

A more detailed explanation of these cultural differences was given by the *Economist* magazine,[‡] in an article entitled "Better Than

*Among the names most often associated with this research are Christoph Bartneck in the Netherlands at Eindoven University of Technology, Cynthia Brezeal at MIT, Lola Cañamaro at the University of Hertfordshire in England, and Sara Kiesler and Illah Nourbakhsh at Carnegie Mellon University.
[†]September 1, 2004.
[‡]December 20, 2005.

People," which explained "why the Japanese want their robots to act more like humans." The article focuses on how these cultural differences affect robotics development in Japan. The reasons are partly economic—the huge growth predicted for the sale of service robots (to $10 billion) by the year 2015—but also cultural.

It seems that plenty of Japanese really like dealing with robots. Few Japanese have the fear of robots that seems to haunt Westerners. In Western books and movies, robots are often a threat, either because they are manipulated by sinister forces or because something goes horribly wrong with them. By contrast, most Japanese view robots as friendly and benign. Robots like people and can do good. The Japanese are well aware of this cultural divide, and commentators devote lots of attention to explaining it. The two most favored theories, which are assumed to reinforce each other, involve religion and popular culture.

Religion plays a role because Shintoism "is infused with animism: it does not make clear distinctions between inanimate things and organic beings." For this reason the attitude in Japan is to question not why the Japanese like robots but why many Westerners view robots as some kind of threat. And this somewhat benevolent attitude toward robots has been enhanced by their popularity, both in newspaper and magazine cartoons and in films, ever since the launch of Japan's robot cartoon character Tetsuwan Atomu in 1951.

]]]]] Robot Chromosomes

A huge step forward on the path to creating robots with humanlike personalities and emotions has recently been taken by Jong-Hwan Kim* and his team at the Robot Intelligence Technology Laboratory in Daejeon, South Korea, who have been working on the development of successive versions of a robot called HanSaRam. In a 2005 confer-

---

*Jong Hwan-Kim was the originator of the robot soccer competitions that have become enormously popular within the electronics and software communities as an intercollegiate and intercorporate sport.

ence paper, "The Origin of Artificial Species," Kim and his colleagues describe the artificial chromosomes they have developed for robots.

The basis of Kim's idea is that the entire collection of a robot's artificial chromosomes will contain all the information about the robot that corresponds to the information stored in our DNA. Thus Kim's programmed genetic makeup is modeled on human DNA, although instead of being a complex double-helix shape as in a human chromosome, each artificial chromosome is equivalent to a single strand of genetic makeup. In humans the principal functions of genetic makeup are reproduction and evolution, but in robots the makeup can also be used for representing the personality of the robot and can be electronically transferred to other robots.

Kim's approach to robot personality was inspired by the evolutionary biologist Richard Dawkins, whose book *The Selfish Gene* asserts that, "We and other animals are machines created by our genes." Kim draws a parallel between humans and humanoids by proposing that the essence of the origins of an artificial species such as humanoids must be the genetic code for that species. His paper presents the novel concept of the artificial chromosome, which Kim describes as the essence for defining the personality of a robot and the enabler for a robot to pass on its traits to its next generation, just as in human genetic inheritance. Thus the artificial chromosome creates a simulation of evolution for its artificial species.

> If we think in terms of the essence of the creatures, we must consider this the origin of artificial species. That essence is a computer code, which determines a robot's propensity to "feel" happy, sad, angry, sleepy, hungry, or afraid.[7]

Continuing the parallel between humans and humanoids still further, Kim suggests that the main functions of a robot's genetic code are reproduction and evolution and that the code should be designed to represent all the traits and personality components of these artificial creatures. Thus his artificial chromosomes, being a set of computerized representations of a DNA-like code, will enable robots to think, feel, reason, express desire or intention, and could ultimately empower them to reproduce,* to pass on their traits to their offspring, and to evolve as a distinct species.

Kim's team has designed fourteen robot chromosomes in all, six of which are related to the robots' motivation, three to their homeostasis,† and four to their emotions. These chromosomes dictate how robots should respond to various stimuli: avoiding unpleasantness, achieving intimacy and control, satisfying curiosity and greed, preventing boredom, as well as engendering feelings of happiness, sadness, anger, and fear and creating states of fatigue, hunger, drowsiness, and so on, all of which will combine to imbue the robot with "life." Kim's robots will be able to react emotionally to their environment, to learn and make reasoned decisions based on their individual personalities.

For ease of development and testing, Kim's simulated chromosomes have been programmed into a simulated creature—a software robot called Rity, living in a virtual world—that can perceive forty-seven different types of stimuli and is able to respond with seventy-seven different behaviors. As determined by their genetic codes, no two Rity robots react in the same way to their surroundings. Some become bored with their human handlers while others, because they have a different personality, pant and express their "happiness" at the sight of their humans. It's all in their genes! One of the next steps by Kim and his team will be to create the equivalent of the human X and Y chromosomes, conferring on robots their own version of sexual characteristics, including lust. Thus if male and female robots like each other, "they could have their own children."

Kim readily admits one of the principal messages of the movie *I, Robot*—namely, that the feasibility of giving robots their own personalities and emotions might make them a danger to humanity. To counter this he suggests employing artificial chromosomes "to design brilliant but mild-tempered and submissive robots," which is one way to ensure

*For a little more on robot reproduction, see the footnote on page 188.
†Homeostasis is a creature's ability or tendency to maintain internal equilibrium by adjusting its physiological processes.

that we do not become enslaved by our creations as they evolve. Given this elementary precaution, by the time "malebots" and "fembots" are available for general consumption the market will be ready for them.

]]]]] The Ten Factors as Applied to Human-Robot Relationships

We saw in chapter 2 how common it is for people to develop strong feelings of affection, including love, for their pet animals. And in chapter 3 we examined the same phenomenon as it relates to virtual pets such as the Tamagotchi. Now we come to examine the ten principal factors that cause humans to fall in love with humans, as discussed in chapter 1. Let us consider which of these factors might also be important in causing humans to fall in love with robots.

At the outset we should recall the importance of proximity and hence repeated exposure as major factors that contribute to placing people in a situation in which falling in love becomes more likely. In the case of a robot, both proximity and repeated exposure are easy to achieve, subject to the robot's cost. Simply buy a robot and take it home and both of these criteria are instantly satisfied.

In chapter 1 we also discussed Byrne's law, which shows that we are more inclined to like someone when we feel good. The empathetic robot, able to determine what makes a particular human feel good, will therefore have a head start in its attempts to seduce. The robot will do its best to create "feel-good" situations, perhaps by playing one of its human's favorite songs or by switching on the TV when its human's favorite baseball team is playing, and then it will exhibit virtual feelings that mirror those of the human, whether they be feelings of enjoyment when hearing a particular song or cheering on a baseball team.

Another lesson from chapter 1 on the subject of getting someone to fall in love with you was that self-disclosure of intimate details can be a powerful influence in this direction. Robots designed to form friendships and stronger relationships with their users will therefore be programmed to disclose virtual personal and intimate facts about their virtual selves and to elicit similar self-disclosure from humans.

Now to the ten reasons for falling in love. Which of them might have parallels in human-robot relationships, parallels strong enough to lead humans to develop feelings of love for robots?

Similarity

Of the most important similarities referred to in chapter 1, only one of them—coming from a similar family background—is not easy for a robot to imitate convincingly, given that its human will know that the robot was made on an assembly line. But as to the other key similarities, I forsee no problem in replicating them, including the most important of all, similarity of personality. It will be recalled from one of Clifford Nass's experiments, described earlier,* that not only do humans prefer to interact with other *humans* of similar personality, but they also prefer to interact with *computers* that have similar personalities to their own. That finding is of great significance when considering the importance of similarity of personality in the process of falling in love. Attitudes, religious beliefs, personality traits, and social habits—information on all of these can be the subject of a questionnaire to be filled out when a human orders a robot, or it could be acquired by the robot during the course of conversation. Once the robot's memory has acquired all necessary information about its human, the robot will be able to emulate sufficient of the human's stated personality characteristics to create a meaningful level of similarity. And as the robot gets to know its human better, the human's characteristics will be observable by the robot, who can then adjust its own characteristics, molding them to conform to the "design" of its human.

One example of a similarity that will be particularly easy to replicate in robots is a similarity of education, since just about all of the world's knowledge will be available for incorporation into any robot's encyclopedic memory. If a robot discovers through conversation that its human possesses knowledge on a given subject at a given level, its own knowledge of that subject can be adjusted accordingly—it can download more knowledge if necessary, or it can deliberately "forget"

*See the section "Robot Personalities and Their Influence on Relationships," page 132.

certain areas or levels of knowledge in order that its human will not feel intimidated by talking to a veritable brain box. This self-modifying capability will also allow robots to develop an instant interest in whatever are its human's own interests. If the human is an avid train buff, then the robot can instantly become a mine of information about trains; if its human loves Beethoven, the robot can instantly learn to hum some of the composer's melodies; and if the human is a mathematician, the robot will have the reasoning powers necessary to prove the popular mathematical theorems of that time. Not only will robots have extensive knowledge, they will also have the power of reasoning with that knowledge.

Desirable Characteristics of the Other

The key "desirable" characteristics revealed by the research literature are personality and appearance. Just as a robot's personality can be set to bear a measure of similarity to that of its human, so it can be adjusted to conform to whatever personality types its human finds appealing. For a robot, as for a human, having a winning (albeit programmed) personality will be arousing in many respects, including sexually arousing. Again, the choice of a robot's personality could be determined partly prior to purchase by asking appropriate questions in the customer questionnaire, and then, after purchase, the robot's learning skills will soon pick up vibes from its human, vibes that indicate which of its own personality traits are appreciated and which need to be reformed. And when its human, in a fit of pique, shouts at the robot, "I wish you weren't always so goddamn calm," the robot would reprogram itself to be slightly less emotionally stable.

A desirable appearance is even easier to achieve in a robot. The purchase form will ask questions about dimensions and basic physical features, such as height, weight, color of eyes and hair, whether muscular or not, whether circumcised (if appropriate), size of feet, length of legs (and length of penis, in the case of malebots). . . . Then the customer will be led effortlessly through an electronic photo album of faces, with intelligent software being employed to home in quickly on what type of face the purchaser is looking for. The refinement of this process can continue for as long as the purchaser wishes, until the malebot or fembot of his or her desire is shown on the order screen. If it's a pert nose that turns you on, your robot can come with a pert nose. If it's green eyes, they're yours for the asking. By being able to choose all these physical design characteristics, you will be assuring yourself of not only an attractive robot partner but also the anticipation of great sex to come.

Personality and appearance are far from being the most difficult characteristics to design into robots. Synthesizing emotion and personality are active research topics at several universities in the United States and elsewhere,* as well as in some of the robotics laboratories in Japanese consumer-electronics corporations. Creating a physical entity in a humanlike form that is pleasing to the eye is relatively straightforward, and the Repliee Q1 robot demonstrated in Japan in

THE REPLIEE Q1 ROBOT WITH HER DESIGNER, HIROSHI ISHIGURO.

*See the section "Designing Robot Personalities," pages 136–40.

2006 is perhaps the first example. By 2010, I would expect attractive-looking female robots and handsome-looking males to be the norm rather than the exception, all with interesting and pleasant (though somewhat unsophisticated) personalities.

Reciprocal Liking

Reciprocity of love is an important factor in engendering love—it is more likely for Peter to fall in love with Mary if Peter already knows that Mary loves him. So the robot who simulates demonstrations of love for its human will further encourage the human to develop feelings of love for the robot.

Reciprocal liking is another attribute that will be easy to replicate in robots. The robot will exhibit enthusiasm for being in its owner's presence and for its owner's appearance and personality. After an appropriate getting-to-know-you period, it will whisper, "I love you, my darling." It will caress its human and act in other ways consistent with human loving. These behavior patterns will convince its human that the robot loves them.

Any discussion of reciprocal liking with respect to robots will inevitably suggest questions such as "Does my robot really like me?" This is an important question, but a difficult one to answer from a philosophical perspective. What does "really" mean in general, and particularly in this context? I believe that Alan Turing answered all such questions with his attitude toward intelligence in machines—if it appears to be intelligent, then we should assume that it is intelligent. So it is with emotional feelings. If a robot appears to like you, if it behaves in every way as though it does like you, then you can safely assume that it does indeed like you, partly because there is no evidence to the contrary! The idea that a robot could like you might at first seem a little creepy, but if that robot's behavior is completely consistent with it liking you, then why should you doubt it?

Social Influences

With time, social influences undergo huge change. What was considered a social aberration fifty years ago or less might now be very much the norm. One important example of this is the tendency in certain cultures for young people to be strongly encouraged to marry within their own culture. Not only are there fewer influences on marital choice nowadays, from parents, peers, and society in general, but there is more resistance from young people to be molded into marital relationships dictated by their cultural and social backgrounds. Attitudes to robots will also change with time—now they are our toys and items of some curiosity; before long the curiosity will start to diminish and robots will make the transition from being our playthings to being our companions, and then our friends, and then our loved ones. The more accepted robots become as our partners, the less prejudice there will be from society against the notion of human-robot relationships, leading more people to find it acceptable to take robots as their friends, lovers, and partners.

Filling Needs

If a robot appreciates the needs of its human, it will be able to adapt its behavior accordingly, satisfying those needs. This includes those relationships in which the human's needs relate to intimacy, even to sex, as explained in part two of this book. One can reasonably argue that a robot will be better equipped than a human partner to satisfy the needs of its human, simply because a robot will be better at recognizing those needs, more knowledgeable about how to deal with them, and lacking any selfishness or inhibitions that might, in another human being, militate against a caring, loving approach to whatever gives rise to those needs.

Arousal/Unusualness

This factor depends for its existence on the *situation* in which a human and the potential love object initially find themselves together, and not on the love object itself. The arousal stimulus is external to the couple. As a result there would appear to be no difference between the effect of a particular arousal stimulus on someone in the presence of another human and the effect of that same arousal stimulus on that same someone in the presence of a robot. In both cases the stimulated human will find the situation arousing, possibly even to the extent that it might make the human feel more attracted to the robot than to another human under the same circumstances. After all, in a situation that appears dangerous, would not a robot be more likely than a human to be able to eliminate or mitigate the danger?

Specific Cues
Absolutely no problem! After a trial-and-error session at the robot shop, you will be able to identify exactly what type of voice you would like in your robot, which bodily fragrances turn you on, and all the other physical characteristics that could act as cues to engender love for your robot at first sight.

Readiness for Entering a Relationship
As in the case of arousal, with this feature it is one's situation that gives rise to the affectionate feelings. If you've just been dumped by your partner and are looking for a flirtation or a fling to redeem your self-esteem, your robot can be right there ready for all eventualities, with no need for speed-dating sessions or for placing an ad in the lonely hearts columns.

Isolation from Others
This is yet another factor where the circumstance dictates what happens. If you have a robot at home, you will be likely to spend considerable time in isolation with it—as much time as you wish.

Mystery
Robots are already something of a mystery to most people. Imagine how much more of a mystery they will become as their mental facilities and emotional capacities are expanded as a result of artificial-intelligence research. This is not to say that robots should be "perfect." By having different levels of performance that can be set or can self-adapt to suit those with whom a robot interacts, the behavior and performance of the

robot can be endowed with humanlike imperfections, giving the user a sense of superiority when that is needed to benefit the relationship. The element of mystery, like variety, will be the spice of life in human-robot relationships.

]]]]] What Does This Comparison Prove?

I submit that each and every one of the main factors that psychologists have found to cause humans to fall in love with humans can almost equally apply to cause humans to fall in love with robots. The logical conclusion, therefore, is that unless one has a prejudice against robots, and unless one fears social embarrassment as a result of choosing a robot partner, the concept that humans will fall in love with robots is a perfectly reasonable one to entertain. It is possible that at first it might only be the twenty-first-century equivalents of Sherry Turkle's 1980s computer hackers* who fall in love with robots, the latter-day versions of the young man who'd "tried out" having girlfriends but preferred to relate to computers. Yet robots in a human guise will be far more tempting as companions and as someone to love than were computers to Turkle's generation of hackers. And even if the computer geeks *are* the first to explore love with robots, I believe that curiosity, if nothing else, will prompt just about every sector of society to explore these new relationship possibilities as soon as they are available. What we cannot really imagine at the present time is what loving a robot will mean to us or how it might feel. Some humans might feel that a certain fragility is missing in their robot relationship, relative to a human-human relationship, but that fragility, that transient aspect of human-human relationships, as with so much else in robotics, will be capable of simulation. I do not expect this to be one of the easier tasks facing AI researchers during the next few decades, but I am convinced that they will solve it.

*See the section "Attachment and Relationships with Objects," page 65.

]]]]] Robot Fidelity, Passion, and the Intensity of Robot Love

For the benefit of most cultures, robots should be faithful to their owner/partner—what we might call robot fidelity.* Robots will be able to fall in love with other robots and with other humans apart from their owner, possibly giving rise to jealousy unless the owner is actually turned on by having an unfaithful partner. Problems of this type can, of course, be obviated, simply by programming your robot with a "completely faithful" persona or an "often unfaithful" one, according to your wishes. How different life would be for many couples if the possibility of infidelity simply did not exist. But, in contrast, while the infidelity of one's robot might be something to be avoided by careful programming, the possibility equally exists for humans to have multiple robot partners, with different physical characteristics and even different personalities. The robots will simply have their "jealousy" parameters set to zero.

Being able to set one's robot to any required level of fidelity will be but one feature of robot design. It will also be appealing to be able to set the love-intensity level and the passion level of your robot to suit your desires. Your robot will arrive from the factory with these parameters set as you specified, but it will always be possible to ask for more ardor, more passion, or less, according to your mood and energy level. And at some point it will not even be necessary to ask, because your robot will, through its relationship with you, have learned to read your moods and desires and to act accordingly.

]]]]] Marrying a Robot

For many of the readers of this book, any discussion on the history or current status of the institution of marriage will take place within the somewhat conservative confines of traditional Judeo-Christian think-

---

*The robot's preset parameters will doubtless include a "polygamy" option, to cater to those religions and cultures in which monogamous relationships are not the norm.

ing and attitudes and those of some of the other major world religions. Within these confines, marriage can only be the union of one man with one woman, a union intended to last for life, a union that usually has as one of its principal goals the creation of children. Yet this view of marriage is not the only view, because there are and long have been cultures within which marriage is viewed very differently. One of the most obvious examples of such differences is that between monogamy, one of the fundamental tenets of marriage in Western society, and polygamy, which is and has been the norm in many other cultures, including tribes in Africa, North and South America, and Asia, and a bedrock of religions such as Mormonism and Islam.* Surely if we are to enter a balanced debate on the history, the current state, or the future of marriage, our discussions should take into account all cultures, their customs, and how they regard marriage. Why should any of us assume that our own attitudes are inevitably the only correct ones and that cultures other than our own are in some way wrong?

America is perhaps the best example in the world of a mixture of races, religions, and cultures that is, precisely *because of* its mix, fast becoming a society in which the tolerance and acceptance of nontraditional customs and ideas create the very basis of society as it evolves. In such a society, if it is to evolve and thrive harmoniously, such acceptance is an essential moral prerequisite. Sometimes we must accept that it is our own views that might be inappropriate, possibly because they are outmoded, and that the more radical, more modern views of others are more suitable for the times in which we live and for the future. This phenomenon, whereby changes in opinion lead to massive social change, has been seen in recent decades with attitudes to homosexual relationships.[†]

---

*The *Ethnographic Atlas* has data on 1,231 societies studied during the period 1960–80, of which only 186 were monogamous societies, while 453 had occasional polygyny (in which a man has more than one female sexual partner simultaneously), 588 had more frequent polygyny, and 4 had polyandry (in which a woman has more than one male sexual partner simultaneously). Since the nonmonogamous societies are in general much smaller than the monogamous ones in terms of population, these statistics do not indicate that monogamy is the status of the minority of the world's population. Far from it.
[†]See chapter 8.

The trend toward the toleration and acceptance of same-sex marriages is but one aspect of the changing face and meaning of marriage. The November-December 2004 issue of *Harvard Magazine* published a highly charged essay, "The Future of Marriage," by Harbour Fraser Hodder,[8] which, although primarily intending to examine how changes in demographics, economics, and laws have altered the meaning of marriage in America, actually makes a number of points that can also be used to support the prediction that marriage to robots will by mid-century raise no more eyebrows than same-sex marriages and civil unions do today. One such point is based on the observation by Nancy Cott, a Harvard professor of American history, that "marriage itself has therefore come in for a broad reassessment."[9]

The reassessment to which Cott refers is that due to the polarizing views of the advocates of same-sex marriage and their "family values"–oriented opponents. Cott explains that "as same-sex couples line up for marriage licenses at courthouses across Massachusetts, opponents predict the death of marriage itself. One side sees tragedy in the making, the other wants to rewrite the script entirely."

It is my belief that marriage to robots will be one of the by-products of the rewriting of the script, a belief rooted in the type of argument employed by those judges who have ruled in support of same-sex marriage. In 1998, for example, in a superior court ruling in Alaska, Judge Peter Michalski called the right to choose one's life partner constitutionally "fundamental,"[10] a privacy right that ought to receive protection whatever its outcome, even a partner of the same sex. "Government intrusion into the choice of a life partner encroaches on the intimate personal decisions of the individual. . . . The relevant question is not whether same-sex marriage is so rooted in our traditions that it is a fundamental right, but whether the freedom to choose one's own life partner is so rooted in our traditions." Michalski's 1998 ruling and many since then have pointed the way not only to a liberalizing of the legislature's attitude to same-sex marriage but also to a strengthening of the attitude toward the right to choose.

The controversy over same-sex marriage is not the only reason why attitudes to marriage in America have undergone dramatic change.

Cott mentions how women's legal identities and their property used to be subsumed into those of their husbands, and we should not forget that in the past, wives were sometimes themselves regarded as the property of their husbands. These issues of unequal ownership have been erased with time, but the subject of ownership seems likely to reappear, though in a completely equal guise, when humans of either sex acquire and thereby own robots that act as their lovers and their spouses.

Cott also touches on another important and relevant change in the history of marriage in the United States, "the dissolution of marital prohibitions based on race." Even though such unions were previously far from unknown, it was not until 1967 that interracial marriages were ruled to be legal in the United States, when the U.S. Supreme Court overruled the sixteen states that still at that time considered marriage across the color line to be void or criminal. The statistics for interracial marriage have since given proof to the overwhelming need for that change: The number of marriages in the United States between African-Americans and Caucasians rose from 51,000 in 1960 to more than 440,000 in 2001.

Same-sex marriage, ownership of a wife and her property, and interracial marriage are but a few of the most significant changes that are apparent from a study of the history of marriage in the United States. Other major changes include an acceptance of the fact that marriage is not necessarily for life, as evidenced by the 50-percent-plus divorce rate in the United States, and the increasing proportion of couples who opt not to have children. All these and other changes of attitude to marriage lead us to the conclusion, succinctly enunciated by Nancy Cott, that "change is characteristic of marriage. It's not a static institution. . . . People can cohabit without great social disapproval; they can live in multigenerational families; there are scenes of group living; there are gay unions or civil unions. There is a greater variety of household forms that are approved and accepted, or at least tolerated. . . ."

Social change is happening faster now than it did two hundred, one hundred, or even fifty years ago, with the result that change in the meaning and purpose of marriage is also happening faster than ever before, and the rate of such change seems certain to accelerate. Chapter 8 provides a relevant example—it is an analysis of how our sexual mores and attitudes have changed over time. In the case of marriage, it seems eminently reasonable to assume that changes in the approval, acceptability, and tolerance of different ideas and new forms of marital relationship will take place over periods no longer than the few decades that were needed to make interracial marriage and same-sex marriage socially acceptable to many and legally acceptable to the state. Cott points out that in the late twentieth century, marriage moved "towards the spouses *themselves* defining what the appropriate marital role or preference is." This newfound freedom for couples to define their respective roles within their marriages now extends into the realm of legal agreement. Elisabeth Bartholet, holder of the Wasserstein Public Interest Chair in Law at Harvard, observes that the legal context of marriage has shifted from one in which the state has "enormous control over marriage" to one where people write "the terms of their own marriage" and are "allowed to have pre-marital contracts."[11] Furthermore, Bartholet comments that the trend of recognizing de facto relationships means that "if you look like a family, feel, smell like a family—you cook meals together, share bank accounts—then you *are* a family for the purpose of the law."

In summary, marriage is changing at such a rate that there appear to be ever-increasing levels of acceptance and tolerance of how any given couple wishes to conduct their lives together. And as part of the right to choose will come the right to choose one's spouse, even a robot spouse. By the time that today's infants are entering matrimony, many of them will be deciding for themselves almost all the rules and laws that are to govern their unions.* By the time *their* children are ready for marriage, around the middle of this century, I believe that such a freedom of decision will be almost universally exercised.

---

*One exception that I do not believe will be eroded, and for very good reasons, is the issue of consent. In my view it should always be an essential prerequisite that the partners in a marriage should agree to it and should be legally considered competent to make such an agreement.

How, then, will today's children and their children make use of their own generations' newfound freedom of marital choice? In attempting to answer this question, we first consider the main criteria employed in the choice of marriage partner. Elaine Hatfield and Susan Sprecher have examined preferences in marital partners in three different cultures—the United States, Russia, and Japan—in preparation for which they selected twelve criteria after studying several other lists of reasons for mate selection from the psychology literature. A total of 1,519 college students took part in their survey (634 men and 885 women), in which they were asked to rate each of the twelve criteria on a scale from 1 (unimportant) to 5 (essential). The results given in Table 1 indicate that of the twelve criteria, only the seventh-ranked—"being ambitious"—and the three lowest-ranked characteristics could reasonably be argued to be inappropriate descriptors for the robots of the next few decades. All six of the top-ranked characteristics will be demonstrable by robots within that time frame, and as for being physically attractive and skilled as a lover, these characteristics will in my opinion be among the first to be demonstrated with some measure of success.

TRAIT	MEAN RATING (OUT OF 5)
Kind and understanding	4.38
Has sense of humor	3.91
Expressive and open	3.81
Intelligent	3.73
Good conversationalist	3.72
Outgoing and sociable	3.47
Ambitious	3.36
Physically attractive	3.27
Skill as a lover	3.17
Shows potential for success	2.95
Money, status, and position	2.50
Athletic	2.50

Ratings of the Mate Selection Traits[12]

TABLE 1

With the freedom for couples to define the parameters of their own marriages will also come the freedom for the individual to define what he or she intends his or her own marriage to mean. Seeking a suitable human spouse might then become not only an exercise in matching interests, personalities, and the various other factors that we know to influence the falling-in-love process but also a search for someone who has used this same freedom of choice as to the meaning, rules, and purpose of marriage to create a model that matches one's own. This relaxation of the constraints that used to provide a stable basis for the rules and expectations of marriage might therefore make it more difficult to find a spouse, since different potential spouses will be looking to play according to different sets of rules. For this reason one of the factors that I believe will contribute to the popularity of the idea of marrying a robot is the avoidance of the difficulty of finding a human partner with matching views on marriage—your robot will be programmed with views that complement your own.

Even more relevant to the practice of marriage to robots will be the question "To what extent will the new freedoms of choice regarding marriage extend to a choice of who (or what) people will legally be allowed to marry?" The United States has already seen some major changes in this respect, as interracial marriage has shifted from illegal to legal and many people's minds and hearts are now open to the possibility of same-sex marriage. And in 2005 the Netherlands hosted a ceremony of a civil union involving three partners—a man and his two "wives"—when Victor de Bruijn, aged forty-six, from Roosendaal, "married" both Bianca (thirty-one) and Mirjam (thirty-five) in a ceremony performed before a notary who duly registered their civil union.

What novel form of civil union will be next? In future decades the sciences of creating prosthetic limbs and artificial hearts and other organs will continue to develop with accelerating pace, perhaps even adding artificial brains to the ever-growing list of body parts that surgeons can replace. The Norwegian philosopher Morten Søby discusses this trend in terms of the manner and extent to which it more and more reduces the distinction between man and machine and "becomes an element in the great story of evolution and development of civilization."[13] Writing about what prosthesis offers for the future, Søby explains that:

> More and more artificial parts are added to the body—the result being a more *artificial* body. Research is being carried out with neural interfaces to develop auditory and visual prostheses, functional neuromuscular stimulants and prosthesis control through implanted neural systems, etc. Biosociological research into complex self-generating and self-referral systems is another example. Information technology and virtualization not only occupy man, nature and culture but are also about to outdate the genre of science fiction.

And to emphasize the point, Søby quotes other prominent philosophers: Paul Virilio in *The Art of the Motor,* who argues that "the basic distinction between Man and machine no longer applies. Both biological research and computer technology question the absolute difference between living machine and dead matter";[14] and Donna Haraway's 1985 essay "A Manifesto for Cyborgs," in which she asserts that "late-twentieth-century machines have made thoroughly ambiguous the difference between natural and artificial, mind and body, self-developing and externally-designed, and many other distinctions that used to apply to organisms and machines."[15]

Thus with artificial limbs, organs and just about everything else body-related blurring the boundaries between real life and virtual life, it is appropriate to ask what impediments need to be lifted to make marriage between human and robot legally and socially acceptable. Right now there is no legal impediment to keep someone with an artificial leg from marrying, nor against someone with two artificial legs, or all four artificial limbs, or an artificial heart. . . . Where and why should society draw the line? Can we reasonably argue that it should be legally acceptable to marry someone 20 percent of whose body is made up of artificial limbs and organs, but that if the proportion were to rise to 21 percent, then such a union should be illegal? What logic dictates that a partner who is half natural and half artificial should be an acceptable marriage candidate but that a three-quarters, or 90-percent, or 100-percent artificial partner should not? Here lies a difficulty for the lawmakers of the future, those who are given the responsibility of drafting changes designed to bring the law up to date. As robots become increasingly sophisticated, as people have them in their homes as companions, when people have sex with them and fall in love with them, so it will become appropriate for those lawmakers to paraphrase Elisabeth Bartholet's argument thus: "If your robot looks like a partner, feels, smells like a partner—you cook meals together, share bank accounts—then you *are* partners for the purposes of the law." And as to the question of a robot's being legally able to consent to its marriage, if it says that it consents and behaves in every way consistent with being a consenting adult, then it *does* consent.

Finally, there are those who would ask, "Why marry?" when discussing human-robot relationships, by which they would mean, "Why would anyone want to marry *any* robot?"—as opposed to why marry a particular robot. Two of the most commonly given reasons as to why people marry are love and companionship. Part one of this book has, I hope, convinced the reader that loving a robot will come to be viewed as a perfectly normal emotional experience and that before very long, robots will be regarded by many as interesting, entertaining, and stimulating companions. If these two reasons for getting married, love and companionship, are the foundation for so many millions of marriages between human couples, why should the same reasons not provide a valid basis for the decision to marry a robot?

]]]]] Some Aspects of the Physical Design of Robots

The eventual acceptance of robots as sentient beings, worthy of our friendship, our love, and our respect will be greatly facilitated by the physical design and construction of robots whose appearance matches our notions of friendliness. Masahiro Mori, head of the robotics department at Tokyo University, was one of the first roboticists to suggest that a robot with a humanlike appearance will be apt to engender feelings of familiarity and affection from humans. This view is borne out by a study based on one of the first controlled experiments to examine the effect of a humanoid robot's appearance on people's responses, with a machine-like robot used as a comparison. The study suggests that people may be more willing to share responsibility with a humanoid as compared with robots that are less humanlike and more machinelike. And if the physical design of a robot creates an appearance in the human image, the robot's physical actions and movements will provide immediate and easily comprehensible social cues, thereby enhancing a human's perception of any interaction with the robot and making it easier for the human to engage with it socially. If, for example, the human swears at the robot, it could stick out its tongue as a gesture of complaint. But if the robot did not have a tongue to stick out, it would not be able to convey its feelings in this humanlike way, while if the robot's tongue were not designed into its mouth but instead were located on the lower part of one of its legs, perhaps the action of sticking out its tongue might not have the same effect on the human.

Even though a robot's appearance brings nothing to bear on its intellectual capabilities, it has been shown by psychologists that in general we prefer to interact with robots with whom we find it easy to identify, as compared to robots whose appearance is strikingly nonhuman.* But there is still a way to go before humanoids are as physically appealing as Stepford wives and their malebot counterparts. Although they are technically remarkable for their time and great fun to watch, the robots of today are not exactly Mr. Handsome or Ms. Beautiful, nor are they as cuddly as pet cats, dogs, rabbits, or Furbies. The Carnegie Mellon University robot, Grace, who attended an academic conference in Canada in 2002, managed to find its way around the conference building well enough to register for the conference, reach a lecture room by itself (asking for directions only when necessary), and deliver a talk on how it worked. But Grace did not look at all humanlike or even animal-like. Its "face" was an image displayed on a com-

*See the discussion on similarity in chapter 1 (page 38).

puter screen that formed the top part of its construction, while the remainder of its body was a mass of metal parts, electronics, wheels, and much of the other paraphernalia one would expect to find in an engineering laboratory. So although Grace performed admirably and with a certain measure of physical dexterity (she could navigate her way into an elevator and exit at the correct floor), she was not exactly anyone's idea of a great-looking date.

One might argue that only the capabilities of a robot should matter to us and not its looks, but I believe that looks will matter a lot, a belief that stems partly from an experience I had around the age of ten. The first time I visited Madame Tussauds museum in London, I asked a gentleman dressed in a uniform the way to some particular part of the exhibition, only to realize after a second or two that he was not on the museum staff—he was one of the waxworks. So convincing was the wax janitor's appearance that I'd been fooled into thinking "he" would respond to my question and would know the answer. After all, he looked just as I expected a museum janitor to look. This experience has doubtless been shared by many thousands of the museum's other visitors, and it is a valuable lesson in understanding an important aspect of human-robot relationships. The appearance of a robot will affect how people perceive it, particularly their first impressions, as well as how they interact with it and the development of their relationships with it. If a robot has all the appearances of being human, then we will increasingly adopt an anthropomorphic attitude toward it and find it much easier to accept the robot as being sentient, of being worthy of our affections, leading us to accept it as having character and being alive. Thus the appearance of a robot's head and face are clearly extremely important factors in our initial reactions when meeting it. First impressions do count. This is why it is not sufficient for the Graces of the future to look like electronics laboratories on wheels, or even on awkwardly moving legs. They must walk in a humanlike fashion, and above all they must be appealing in their appearance. Only then will huge numbers of people want them as their friends and lovers.

One year after Grace made her debut as a conference attendee,

David Hanson, a graduate student at the University of Texas at Dallas, demonstrated a lifelike talking head. Its face had soft, flesh-colored, artificial skin made of an elastic, flexible polymer developed by Hanson especially for this purpose. The face on Hanson's artificial head had finely sculpted cheekbones and big blue eyes. When connected to a computer the head could smile, it could frown, it could sneer, and its brow could develop furrows to give a worried look. Equally, the robot could turn its head, and particularly its eyes, toward a human, taking in through its vision system whatever emotional cues the human might be exhibiting and using this information to help it react with appropriate facial expressions. This kind of expressive power will enable robots to interact more easily with humans, using their electronic minds to control their facial expressions and head movements in accordance with whatever emotions the robot wishes to display. It is part of the human mechanism for developing two-way emotional relationships, a mechanism that will be enhanced with the affective technologies described earlier in this chapter.*

The design of the head that Hanson demonstrated was based on that of his blue-eyed girlfriend, Kristen Nelson. In April 2002, he had gone to a bar in the trendy Exposition Park area of Dallas, complete with a pair of calipers, in search of someone whose head would be suitable as a model for what Hanson had in mind. There he saw Kristen, whom he knew casually, and asked her, "Can I make you into a robot?" He did. The movements of Hanson's artificial head are made possible by a collection of twenty-four motors, invisible to the observer, that simulate the actions of most of the muscles in the human face. The motors are driven by two microprocessors, and they employ nylon fishing line to tug the artificial skin when it needs to move. The eyes contain digital cameras to enable the head to see the people who are looking at it and, if required, to imitate their facial expressions, courtesy of its "muscle" motors.

Following its first convincing demonstration and the aura of publicity that surrounded it, the head attracted interest from companies in

*See the section "Robot Recognition of Human Emotions," page 124.

(LEFT TO RIGHT) DAVID HANSON'S ROBOT VERSION OF KRISTEN NELSON'S HEAD, KRISTEN, DAVID.

fields ranging from artificial limbs to sex dolls. And that was in 2003. In the time line for the development of sentient, lovable robots, Hanson's work puts head design ahead of schedule. Add Hanson's artificial head to Grace's body and already the physical appearance of robots will have reached new heights of acceptability. And just as a robot's emotional and intellectual makeup and its face and voice can be selected on an individual basis, so it can be designed with any wished-for physical characteristics, including skin, eye, and hair color; size of genitalia; and sexual orientation.

]]]]] Feel and Touch Technologies

In designing artificial skin for robots, the most important properties will probably not be its appearance and expressiveness but rather its sensing capabilities—feel and touch. From a purely practical perspective, having a well-developed sense of feel will enable a robot to detect changes in its surroundings and move accordingly. But it is the more romantic aspects of feel that concern us here—how a robot can detect a physical expression of love, a caress or a kiss. Though perhaps with different research goals in mind, scientists in Japan, Italy, and the United States are working on high-tech skin development. The sensuous robot will be one of the spin-offs of their research.

At the University of Tokyo, a group led by Takao Someya is developing a synthetic skin, based on the technology for printing enormous numbers of flexible, low-cost pressure sensors on a large area of the skin material. Meanwhile in Italy, at the University of Pisa, Danilo de Rossi and his team are making skin using artificial silicone, which has the properties of elasticity (human skin stretches if pulled) and sensitivity to pressure. And in the United States, scientists at NASA are employing infrared sensors embedded in a flexible plastic covering—the sensors detect an object as the robot touches it and then send a signal to the robot's computer, its "brain," corresponding to the size, shape, and feel of the object.

The different types of sensor and the different skin materials being investigated by these groups reflect that the study of artificial-skin technology is still in its infancy and there is not yet a consensus as to what materials and technologies make for the best artificial skin. Future artificial-skin materials are likely to be more tactile and to provide even more sensors to afford greater sensitivity, but from the perspective of skin as an important component of a robot love object or sex object, it is hardly important what types of sensors are being used, or how many. What *is* important is that robots will be able to feel and recognize the touch and caress of an affectionate human, to know when their human is making the first physical overtures of passionate, romantic love. Similarly, a delicate sense of touch will be needed *by* a gentle robot lover, able to return its human's tender caresses and initiate its own. Scientists at the Polytechnic University of Cartagena in Spain have created a sensitive robotic finger that can feel the weight of pressure it is exerting and adjust the energy it uses accordingly, allowing a robot to caress its human partner with the sensitivity of a virtuoso lover.

]]]]] Smell and Taste Technologies

One novel technology that will contribute to a robot's physical appeal is smell synthesis. The right kind of bodily fragrance can act as a powerful attraction and aphrodisiac, and not necessarily the kind of scent that comes in small bottles with big price tags. Instead the idea is to create electronically any smell to order. Just as your stereo speakers play out digitally stored music, so its smell equivalent will spray out the digitally stored smells generated by this technology. Your robot can exude a favorite perfume or a realistic counterfeit of your (human) loved one's body fragrance, or even a body fragrance of its own that has been designed to appeal to you and to cater to your hormones and your personal desires.

The early attempts at bringing smell technology to the market were not exactly a great success. Despite serious investment, reportedly $20 million in one company alone,* the sweet smell of success eluded the pioneers in this field. By 2005, however, a new generation of digital-smells companies were racing to be the first to launch viable smell-creation technology,† and technologies very similar to those employed in the generation of smells to order can also be employed in the creation of artificial flavors that taste just like the real thing.

The fascinating aspect of this technology, from the perspectives of love and sex, lies in the creation of scents that can set a partner's hormones running. These sense technologies will provide some of the foundation for the amorous and sexual attraction that humans will feel for robots. Sex usually involves several senses simultaneously: We enjoy the sight of our loved one, we enjoy the sound of their voice, the feeling of their skin when we caress it and the feeling on ours when we are touched, we enjoy their smell and their taste. All of these senses heighten our erotic arousal, and all of their corresponding technologies can be designed into robots to make them both alluring and responsive.

]]]]] Robot Behaviors

An important facet of designing robots that promote satisfactory relationships with humans (satisfactory from the human point of view) is

---

*DigiScents.
†Trisenx (www.trisenx.com), the French company Exhalia (www.exhalia.com), SAV Products of California (www.savproducts.com), and an (as-yet-anonymous) Israeli company all appear to have similar technology.

an analysis of the extent to which the robot needs to behave in a sociable way with humans in different types of situation. If, in a particular situation, a robot exhibits none of the normal human characteristics of emotion, it will probably appear to be insensitive, indifferent, even cold or downright rude. Solving this problem is not that simple. There might be some people—some nationalities, some age groups, or one of the sexes—who do not perceive a robot to be any of these things in the given situation, simply because of their cultural, educational, or social background. What is cold, rude, or uncouth to one group in society might appear to be completely normal, acceptable, even friendly to another group. A sociable robot that has emotional intelligence will therefore need to be able to make this distinction, to decide how to behave with different people in the same situation in order to be perceived as sociable by all of them. (Robots will be programmed to want to be liked by everyone, just as you and I do.)

Other factors that might affect the appropriate way for a robot to behave include where the human-robot interaction is taking place. Is it in the home, where a more overtly friendly behavior by the robot would be appropriate? Or is it at work, where the human might be the robot's boss (or vice versa), and therefore a more overtly respectful attitude would be required of the robot (or the human)? Robots will need to be endowed with many "rules" of sociability for all sorts of situations and contexts, and this rule set can be expanded through the use of learning technologies. If a robot acts in a manner that appears rude to a human, the robot can simply be told, "That is rude," whereupon, like a well-brought-up child, the robot can learn to improve its manners and behavior.

An interesting question here is whether robots should merely be designed to imitate human sociability traits or whether they should be taught to go further and create sociability traits of their own, traits that are atypical of humans but can nevertheless be appreciated by humans. To do so would be a form of creativity, possibly no more difficult to program than the task of composing "Mozart's" Forty-second Symphony or painting a canvas that can sell in an art gallery for thousands of dollars—tasks that have already been accomplished by AI researchers.*

At the ATR Intelligent Robotics and Communication Laboratories in Kyoto, a robot called Robovie has been developed as a test bed for ideas in robot-human communication. Robovie has a humanlike body that is only four feet tall, so as not to be overly intimidating to the humans outside the laboratory with whom it comes into contact from time to time. Robovie has two arms, two eyes, and a system of three wheels to enable it to move around. (Legs are not yet considered a necessity for Robovie's principal sphere of activity, which is communication with humans rather than tasks involving movement.) Robovie has an artificial skin, to which have been attached various sensors, sixteen of them, made from pressure-sensitive rubber. It can speak, it can hear and recognize human speech, and it can charge its own batteries when necessary.

Robovie's developers believe that there is a strong correlation between the number of appropriate behaviors a robot can generate and how intelligent it appears to be. The more often a robot can behave in what is perceived to be an appropriate manner, the more highly will its intelligence be regarded. The scientists developing Robovie plan to continue to develop new behavior patterns until Robovie has advanced to the point where it is much more lifelike than a simple automaton. Part of this progress will come from the robot's tendency to initiate interaction with a human user, rather than merely being reactive. You and I don't always wait until we are spoken to before we say something, so why should a robot? You and I don't always wait until someone stretches out their hand to us and says, "Hi. Nice to meet you." Nor should a robot. Robovie will in appropriate circumstances shake hands with you; hug you; greet, kiss, and converse with you; play simple

*David Cope, at the University of California at Santa Cruz, has developed a program called EMI (Experiments in Musical Intelligence) that composes music in the style of Mozart, Chopin, or Scott Joplin, among others. And another California professor, Harold Cohen, has developed AARON, a drawing and painting program whose talents include controlling a robot that can wield paintbrushes with skill and even knows when the paint pot is running dry.

games such as rock-scissors-paper; and sing to you. And these are just some of the behavior patterns it had been taught up to mid-2004.

Robovie's arms, eyes, and head also contribute to the robot's ability to interact with humans and to how they perceive it, partly because of the importance of eye contact in the development of human relationships and therefore in the creation of empathetic robots. We humans greatly increase our understanding of what others are saying to us, the subtext as well as the words themselves, when we establish eye contact and observe a speaker's body gestures. Research has repeatedly shown that during a conversation humans become immediately aware of the relative position of their own body and that of the person to whom they are speaking—the body language improves the communication. This explains the tendency for Japanese roboticists to build human-shaped robots, endowing them with effective communication skills and employing the results of research from cognitive science to create more natural communication between robot and human.

Experiments with a group of twenty-six university students showed that Robovie exhibits a high level of performance when interacting with humans, while the students generally behaved as though they were interacting with a human child, many of them maintaining eye contact with the robot for more than half the duration of the experiment. Some of the students even joined in with the robot in its exercise routines, moving their own arms in time with the robot's movements. The natural appearance of the students' interactions in the experiment was attributed to the humanlike appearance and behavior of the robot.

]]]]] Humanoid Robots—from the Laboratory to the Home

The development of humanoid robots has thus far been a long and slow process. The first serious development of humanoids began at the School of Science and Engineering at Waseda University in Japan, with the commencement of the WABOT project in 1970. The first full-scale humanlike robot, WABOT-1, was completed in 1973. It could talk (in Japanese), it could measure distances, it could walk, and it was able to grip and carry objects with hands that incorporated tactile sensors to allow the robot to feel what it was carrying. It also had an artificial mouth, ears, and eyes.

In 1984 came the musician robot WABOT-2, designed to play a keyboard instrument. This task was chosen by the Waseda engineers as one that requires humanlike intelligence and dexterity. WABOT-2 could read a musical score, play tunes of average difficulty on an electronic organ, and accompany someone who was singing a song.

The most dramatic development thus far in the Waseda project started in 1986: creating a robot that can walk like a human. Well, almost. Its feet edge slowly and deliberately forward, and even after twenty years' research it is not yet able to qualify for the walking championship in the Olympic Games. But it has long been able to climb up and down stairs and inclines, it can set its own gait so as to be able to move on rough terrain and avoid obstacles, and it can walk on uneven surfaces.

]]]]] The March of the Humanoids

Once upon a time, before the advent of the PC, computers were so expensive that they were rarely found outside the confines of government, big business, and academia. Reasons for this expense included the high cost of powerful processing units—the "electronic brains" that enabled the computers to compute—and of the computer memories that had to be employed to store the programs and their data. All this changed in the late 1970s, when inexpensive microprocessors became available, devices that cost a few dollars but could perform calculations and the electronic manipulations of data that only a few years earlier would have required a "mainframe" computer.* Suddenly there were computers in the home, such as the Commodore PET and the Sinclair Spectrum, inveigling themselves into people's daily lives.

---

*"Mainframe" was the term used for large, powerful computers that often served many connected terminals and were usually installed at large organizations.

Androids have not yet reached that level of integration into our society, but their day is fast approaching.

Robots are not yet just like us, obviously. They behave in most respects in what we currently refer to as "robotlike" or "robotic" ways. One physical manifestation of this is how biped robots walk, slowly and deliberately moving their feet, making it obvious to the observer that they're thinking about every step. Even the most advanced android robots today move in this extremely slow and deliberate manner.* Similarly, the best of today's conversational software can be recognized as artificial by just about all the judges at the annual computer-conversation competitions. So as yet we cannot fairly describe our robots as being sociable, because to be considered sociable they would first need to be more humanlike. But that will come. When robots are perceived as making their own decisions, people's perceptions of them—as solely tools for mowing the lawn and other domestic tasks—will change. And just as the day will arrive when, all of a sudden, robots are sufficiently humanlike to be considered for the epithet "sociable," so the day will also come when robots are sufficiently sociable, in human terms, to be considered as candidates for our deepest affections.

Why do I believe that the necessary change in thinking will take place among a wide body of the population, a change sufficiently dramatic to alter people's perception of robots from that of servants to their being our friends, companions, and more? It is because we have already seen other instances of the process necessary to bring about similar changes in our ideas about the roles of robots. This process requires two components—a change in our social and/or cultural thinking and a significant leap in technological capability.

There are several examples from the twentieth century of major social and cultural changes—particularly those relating to women: their enfranchisement as voters; their role in the home and in parenting, developing from that of dutiful housewives to members of a more equal partnership; their role in the workplace, from filling only the

*An excellent and often updated source on the topic of humanoids is the Web site Historical Android Projects at www.androidworld.com.

more menial jobs to taking on management and executive positions; the advances in female contraception that have given women more choices regarding their lifestyles and careers. Society is also undergoing a change in ideas regarding senior citizens, moving away from the expectation that one works with retirement in mind—and the sooner the better—to what is becoming regarded as a more economically sound model—namely, that later retirement means more earning potential and a lesser financial burden on the state, on one's children, and on inadequate pension schemes. Another change that has become apparent in recent years is in society's view of human appearance, as our concerns over obesity can be seen to lead to cultural expectations regarding the "correct" body size and shape, the result being that many women develop eating disorders while they try to stay (or become) thin. Also more apparent nowadays are cultural changes in individuals, as those who encounter people of other cultures sometimes question the ideas and conventions of their own culture, and change as a result.

Leaps in technology occur frequently. In the case of humanoid robots with the capabilities described in this book, most of the more difficult advances will be in the realm of the robot's software—the computer programs that give it emotions and personality, that enable it to think, to understand what is said to it, to conduct a conversation, to make intelligent deductions and assumptions. These advances will come partly through new techniques in artificial intelligence—in other words, through new programming ideas—and partly because of developments in computer hardware, in the chips or whatever it is that will do the thinking, and in the computer memories that store the massive amounts of information robots will need. We have seen for many years that computing speeds and computer memory sizes increase steadily, year upon year, but the increases we have witnessed during the past two or three decades will pale into insignificance when completely new technologies become mainstream, technologies that go under names such as "optical computing," "quantum computing," "DNA computing," and "molecular computing." So rest assured, the advances in technology needed to create the robots that I describe in this book will indeed come. It is only a matter of time, and technological advances are happening ever faster as time goes on. The more we know about a science, the faster we are able to discover even more about that science and to develop technologies based on this new knowledge.

When we combine significant change in our social and cultural thinking with massive advances in technology, one result is the creation of entire new product categories, products that take advantage of new technologies to implement the ideas that make social change possible. When we have the technology, when we are receptive to the social change, society will move forward in that new direction. Robots as dance partners, for example—in 2005, Tohoku University in Japan demonstrated a dancing robot that can predict the movements of its dance partner, enabling it to follow its partner's lead and to avoid treading on any toes. Another example is robots as university lecturers and public speakers—Hiroshi Ishiguro, from Osaka University's Intelligent Robotics laboratory, has made casts of himself that form the basis for clones that he sends to deliver lectures in his stead. Then there is the robot sales assistant, developed by Fujitsu, that works in a Japanese department store, guiding customers around the store and carrying their shopping. And a receptionist, only twenty inches tall, manufactured by the Business Design laboratory in Nagoya, Japan, that asks visitors their name, can recognize as many as ten different faces, and tells visitors when the person they have come to see is ready to meet them. The examples go on and on, every year coming with its own crop of new applications for robots.* Robot jockeys that ride camels in races, robot butlers . . . And most of them, as you will have realized by now, are developed in Japan.

One non-Japanese product that has been a big commercial hit is the Robosapien android robot, a Chinese-American coproduction. Robosapien was the first affordable humanoid to come on the market. It was a toy designed by Mark Tilden, a former NASA scientist, manufactured in China and incorporating simple forms of some of the tech-

---

*A good way to stay abreast of the latest in robotic achievements and capabilities is to visit the Web site of the American Association of Artificial Intelligence at www
.aaai.org/AITopics/html/robots.html and select the "General Index by Topic to AI in the News."

nologies described in this book. It could exhibit several movement-related capabilities, including using its articulated arms to pick up objects such as cups, socks, pencils, and other small light objects; throwing, dancing, and effecting a few karate moves. The toy reacted to touch and sound signals and had sensors in its feet to enable it to detect and avoid obstacles. It could also walk at two different speeds. Robosapien had personality as well—if it wasn't given any commands for a while, it would go to sleep and start to snore! At the price, around eighty-nine dollars in the United States, Robosapien was a sensation. The first of its kind.

The commercial success of the Robosapien during the second half of 2004, when in Britain alone some 160,000 were sold, was perhaps the first stage of the assimilation process for robots. Robosapien was remarkable mainly for its ability to perambulate, albeit in a typically deliberate and robotic manner. When vision technology is added to enable this toy and others to recognize people and objects, when natural-language-processing and speech-synthesis technologies understand what people say to them and to reply sensibly, when cognitive technologies learn and are able to plan how to solve problems, then robot toys will become part of the family, rather like a new breed of family pet. But instead of requiring feeding, vet bills, and expensive places to stay when you take your vacation, these electronic pets will carry a once-in-a-lifetime cost of a hundred bucks or thereabouts, rechargeable batteries included. In the meantime humanoid robots are somewhat more expensive. Mitsubishi's Wakumaru will look after your house while the family is absent, monitor the health of a sick relative, connect itself to the Internet and sort your e-mails, recognize up to ten faces, understand some ten thousand spoken words (in Japanese), encourage you to visit the gym, and be "convenient for the life of family members."[16] A real deal, at around $14,300.

In concluding the first part of this book, I very much hope that any readers whom I have failed to convince as to the viability of emotional relationships between humans and robots will not close their minds to the possibility but at least be willing to observe without prejudice as advances in robotics and AI arrive thick and fast during the coming years. Deb Levine's stimulating turn-of-the-milennium article "Virtual Attraction: What Rocks Your Boat," makes an excellent case for at least remaining open-minded:

> As time goes on, it will be important for society to recognize the various ways people are interacting intimately as valid and equal. Right now, some relationships, specifically marriage between heterosexual couples, are valued more than others are. As technology enters more people's lives, and we are exposed to a variety of different attractions and relationships, it will be important to recognize and equalize virtual forms of attraction and communication with more traditional face-to-face interactions.[17]

PART TWO

> > > > > > > > > > *Sex with Robots*

Best sex I ever had! I swear to God! This RealDoll feels better than a real
woman! She's fantastic! I love her! This RealDoll is for real, I swear! Better
than a woman! My wife isn't as good as that! May God take away all my
ratings if I'm lying! I'll take a lie detector test! I swear on the life
of my children! I did it and it was fulfilling! I did it and I'm proud of it!
It was great! It was the best sex I ever had! . . . It was fabulous!
I could fall in love with that thing!

—Howard Stern

Introduction to Part Two

Sex with humanlike artifacts is by no means a twenty-first-century concept—in fact, its foundations lie in the myths of ancient Greece. A Cypriot sculptor, King Pygmalion I, made an ivory statue in the form of a woman that was so beautiful he fell in love with it, gave it a name—Galatea—and desired it. So he prayed to Aphrodite, the goddess of love, and one day while Pygmalion was kissing the statue, Aphrodite brought it to life. Pygmalion's kisses were suddenly being reciprocated, and finally he married Galatea. The myth of Pygmalion thus led to the name tag "pygmalionism," for the fetish of sexual attraction to statues.*

In his authoritative 1909 tome *The Sexual Life of Our Time,* Iwan Bloch explains one of the oldest of religiosexual phenomena, the act of "religious prostitution," as a form of pygmalionism. This is an act of sacrifice, made to a deity, most often taking the form of a sacrifice by a woman of her virginity shortly before giving herself to her husband for the first time. The defloration process would sometimes be accomplished with a penis made of ivory, stone, wood, or even iron and sometimes by a form of pygmalionism—intercourse with a statue of the god. As an example of this practice, Bloch describes how a bride at a religious shrine near Goa would be assisted by her friends and relatives in mounting the stone penis of an image of a god, thereby destroying her hymen.

*This fetish also goes under the name "agalmatophilia."

In this religiosexual act, the statue is a representation of a deity, but in the far more common form of pygmalionism the statue substitutes not for a deity but for a living human being. In the brothels of late-nineteenth-century Paris, it was not uncommon for prostitutes to act out a variation on this theme, standing on suitable pedestals as though they were statues and being watched by their clients as they gradually appeared to come to life. Such a scene induced sexual enjoyment in the Parisian pygmalionists, often elderly patrons who no longer had the energy for sex. At about the same time, the French talent for inventing mechanical automata such as Vaucanson's duck and Maillard's swan,* when combined with the legendary French expertise in matters sexual, led to the invention of artificial devices, and even whole artificial bodies, designed to provide substitutes for human genitalia.

Bloch describes how these were employed, to act as surrogate sex partners.[1]

> . . . we may refer to fornicatory acts effected with artificial imitations of the human body, or of individual parts of that body. There

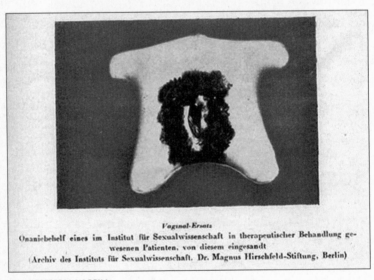

Vaginal-Ersatz
Onaniebehelf eines im Institut für Sexualwissenschaft in therapeutischer Behandlung gewesenen Patienten, von diesem eingesandt
(Archiv des Instituts für Sexualwissenschaft, Dr. Magnus Hirschfeld-Stiftung, Berlin)

AN ARTIFICIAL VAGINA

*See the introduction, page 3.

exist true Vaucansons in this province of pornographic technology, clever mechanics who, from rubber and other plastic materials, prepare entire male or female bodies, which, as *hommes* or *dames de voyage,* subserve fornicatory purposes. More especially are the genital organs represented in a manner true to nature. Even the secretion of Bartholin's glands* is imitated, by means of a "pneumatic tube" filled with oil. Similarly, by means of fluid and suitable apparatus, the ejaculation of the semen is imitated. Such artificial human beings are actually offered for sale in the catalogue of certain manufacturers of "Parisian rubber articles." A more precise account of these "fornicatory dolls" is given by René Schwaeblé ("*Les Détraqués de Paris,*" pages 247–53).

From René Schwaeblé's description of these fornicatory dolls, sold by a "Dr. P" for around three thousand francs, it would appear that they were extremely convincing replicas of the female form.[†] The doctor explained to Schwaeblé:

Every one of them takes at least three months of my work! There's the inner framework which is carefully articulated, there's the hair on the head, the body hair, the teeth, the nails! There's the skin, which has to be given a certain tint, certain contours, a particular pattern of veins. There are the eyes, which need to be given some expression, there's the tongue, and I don't know what else. You won't find a waxwork or a statue, not even the ones created by the greatest masters, that can be compared to my products. The only thing *these* haven't got is the power of speech!

. . .

Unfortunately I can't advertise openly. The police keep interfering in my business, and I have to keep some weird rubber animals around the place, so that I can say I'm a maker of inflatable figures for funfairs!

---

*The glands located on either side of the vaginal orifice that secrete a lubricating mucus.
†This translation of pages 247–53 of Schwaeblé's book is by John Sugden.

Doctor P occasionally had customers who wanted a doll made in the likeness of someone they desired.

It quite often happens that one of those "mad women" falls for a man in the public eye—a politician, a jockey, some hammy actor, or whatever. As she doesn't dare to become his mistress, or can't, she applies to me and asks me to create a doll modelled on her idol.

. . .

Madame X—— lost her husband last year. Two days after his death, she came to me and asked me to craft a doll in the image of the deceased. Didn't she get on my nerves! Every afternoon she would settle herself in my studio and watch me at work, showering me with advice: "Skin more pink *here*! More hair *there*! Lip curling up a little! A more cheerful eye!" When the doll was finished she took it home with her. Since then she's been living with it, she never leaves it. She dresses it in her husband's own clothes, puts it to bed beside her at night, kisses it, caresses it and tells it all sorts of naughty things![2]

With real products available for purchase in fin de siècle France, such as the one described here by Schwaeblé, it is hardly surprising that French fiction of that time made use of fornicatory dolls. Bloch wrote:

The most astonishing thing in this department is an erotic romance *La Femme Endormie,* by Madame B.; Paris, 1899, the love heroine of which is such an artificial doll, which, as the author in the introduction tells us, can be employed for all possible sexual artificialities, without, like a living woman, resisting them in any way. The book is an incredibly intricate and detailed exposition of this idea.[3]

So "shocking" was the content of *La Femme Endormie* that not only did the author feel the need for anonymity, but the book boldly displayed the misinformation that it was printed in Melbourne, in an

Koitus-Ersatz
Entkleidbare, mit allen weiblichen Kleidungsstücken ausgestattete primitive Puppe fast in
Lebensgröße
(Archiv des Instituts für Sexualwissenschaft, Dr. Magnus Hirschfeld-Stiftung, Berlin)

Tab. XX   Zur Sittengeschichte des Pygmalionismus

FORNICATORY DOLLS

attempt to throw off any straitlaced French authorities who might be
seeking to take legal action against the printer or to prevent further
copies from being distributed.

Is it a far cry from titillating nineteenth-century French fiction to
mid-twenty-first-century sexual robots? Part two of this book aims to
convince any skeptics among you that this transition will indeed mate-
rialize.

5 | Why We Enjoy Sex

The idea of sex with robots affects different people in different ways. Some regard the concept as totally outlandish, arguing that only sex with another human being can be a meaningful and enjoyable experience. Some rely on religious objections based upon the idea of sex as being intended solely for procreation. Others are curious as to exactly how a robot would function sexually and how it would feel for the human. Some embrace the idea wholeheartedly and want to know, "Where can I buy one?"

In this and the following three chapters, I hope at least to dispel any suggestions of outlandishness and to present what I believe are compelling arguments to show that sex with robots will become the norm rather than being an oddity. We start by examining sexual relationships between humans. This we do from a graded perspective, though the gradation is not one whose range lies between lousy sex and great sex—rather it transcends a spectrum of categories of sexual partner. At one end of this spectrum is the passionate love of our life. At the other end lies someone whom we do not even know, have never met before the first sexual encounter (which might be the only encounter with this particular sex object), and who has little or no reason to offer any genuine affection before or during sex. I hope that by explaining why people have sex with people across this entire spectrum, even with those at the "bottom" end of the range, I will be able to convince those members of the "totally outlandish" persuasion that for

WHY WE ENJOY SEX

many people sex can be an enjoyable experience even when the sex object is off the bottom of the range altogether, when instead of a human sex partner there is a sexual robot. We start by examining some fundamental aspects of human sexuality—what are our motives for having sex, and why do we enjoy it?

]]]]] Why Do People Make Love (with People)?

Half a century after Freud's 1938 proclamation that pleasure is the goal of sex, psychologists began to analyze methodically the most common reasons for making love. In some of the earliest of those studies, it was found that traditional stereotypes reflected the actuality of the different reasons men and women engage in sex. A study by John DeLamater in 1989 found that twice as many women as men claimed to have been in love with their first sexual partner, while another study found that 95 percent of college women but only 40 percent of college men responded that for them emotional involvement was "always" or "most of the time" a prerequisite for having sex. When researchers asked the specific question "What would be your motives for having sexual intercourse?" women typically gave reasons relating to love, while the answers from men focused much more on the physical pleasure. And when the question was even more focused, inquiring about the subject's most recent sexual encounter, 51 percent of women and 24 percent of men gave reasons connected with love and emotion, while 9 percent of women and 51 percent of men gave answers relating to lust and physical pleasure. These results have generally been confirmed by subsequent research in experimental psychology.

The general drift of this research might seem to suggest that men will be more likely than women to be interested in participating in sex with robots, based on the assumption that men are more likely than women to be willing or indeed eager to satisfy their sexual desires, even without any emotional attachment to their chosen sex object. On the contrary, I believe that eventually women will exhibit every bit as much enthusiasm as men for sexual coupling with robots, but the women's reasons will often be different—men will want the pure physical pleasure of intercourse and orgasm with robots, while most women will want not only a personal demonstration of the robot's virtuoso lovemaking skills but also to feel the robot's virtual love for them.

In 1989, Barbara Leigh used a survey among 580 people taken from 4,000 randomly chosen households in the San Francisco area as the basis for an analysis of seven reasons for having sex (Table 2). Heterosexual participants were asked to rank each of the seven reasons on a scale of 0 to 4, from "not at all important" (scoring 0) to "extremely important" (scoring 4). The highest score from the two groups was 3.7 out of a maximum of 4 for the "pure pleasure" motivation for men, supporting Freud's belief in pleasure as the goal of sex.

| | FREQUENCY FOR | |
REASON	MEN	WOMEN
For pure pleasure	3.7	3.1
To express emotional closeness	3.5	3.6
To please your partner	3.2	2.7
Because your partner wants to	2.8	2.5
To relieve sexual tension	2.5	2.0
To reproduce	1.2	1.2
For conquest	0.6	0.3

TABLE 2

A more recent study by Valerie Hoffman and Ralph Bolton expanded the above list from seven reasons to sixteen. In addition to the factors listed by Leigh in Table 2, they employed factors from two other studies, including one from 1984 of college students in which the participants revealed their reasons for deciding to have sex for the first time with a recent partner. The sixteen reasons in the Hoffman-Bolton study were put to 146 heterosexual men, generally well educated, who were asked to indicate how frequently each reason applied to their sexual encounters (the scale in Table 3 ranges from 0, meaning that it never applies, to 4, meaning it always applies*). Note that the Hoffman-Bolton list does not explicitly include "own pleasure" as a reason, but their published results make it clear that four of their reasons are highly correlated with obtaining pleasure: "to have fun," "to please my partner," "because I want new experiences," and "to reduce tension."

REASON	FREQUENCY
To please my partner	2.80
To express love	2.78
To have fun	2.77
To feel close emotionally	2.49
To feel loved	2.14
Because my partner wants me to	2.11
To reduce tension	2.10
Because I want new experiences	1.99
To avoid boredom	1.32
Because I feel I just have to	1.25
Even when I don't want to	1.11
For conquest	0.92
Because I'm drunk	0.86
To express domination and power	0.85
Because I'm high	0.69
To have children	0.49

TABLE 3

The majority of the motivations listed in both of the above surveys are presented in somewhat egocentric terms, generally indicating something that the respondents want for themselves out of the sexual experience. In contrast, a third study, in 2004, which was based on a survey conducted via the Internet, expressed ten of its eleven proffered motivations in terms of the way the respondents related to and felt about their partners. These were the following: to achieve emotional

---

*In fact Hoffman and Bolton employed the range 1 to 5, but here the numbers have been converted to the scale 0 to 4 for ease of comparison with Leigh's results.

closeness, raising one's self-esteem by increasing the feeling of being desirable and wanted, to nurture and care for the partner, experiencing the partner's power, to obtain approval and reassurance from the partner, to disarm the partner and protect oneself against hostility or the partner's negative moods, to exert power or control over the partner, and to elicit nurturing and caregiving from the partner.

| | FREQUENCY FOR | |
REASON	MEN	WOMEN*
Emotional closeness	2.95	2.85
Physical pleasure	2.93	2.67
Enhance self-esteem	2.68	2.85
Nurture	2.46	2.19
Feel partner's power	2.25	2.19
Reassurance	2.06	1.94
Self-protection	1.94	1.92
Stress reduction	2.10	1.70
Feel one's own power	1.83	1.77
To manipulate the partner	0.83	0.95
To have children	0.69	0.69

*The scores in Table 4 have been converted, for ease of comparison, from the scale of 1 to 9, as used by Deborah Davis and her colleagues, to the scale of 0 to 4 as in Tables 2 and 3. Table 4 is presented in descending order of the mean scores from male and female respondents.

TABLE 4

In all three of these studies, pleasure and emotional closeness were at or very near the top of the respondents' lists, so we shall examine them first when considering why humans might want to have sex with robots. Let us start with pleasure.

The most obvious way in which humans obtain pleasure from sex is through orgasm, and a robot that can give its partner great orgasms on demand will therefore be highly prized as a sexual partner. In chapter 7 we shall discuss the technologies that will most likely be employed in robots for achieving this goal, but in the meantime consider how a simple sex doll, with no electronic brain, no artificial intelligence, and none of the humanlike characteristics that come from these technologies, can help men to achieve great orgasms. In 1997 the popular radio "shock jock" Howard Stern was given the "Celine" model of a RealDoll by its manufacturer, Abyss Creations. Stern tried it out and waxed lyrical on his radio show about the experience, proclaiming it to be "the best sex I ever had! I swear to God!"

This primitive example shows how much pleasure a mere doll can bring its owner, through the simple expedient of being a nonactive partner in an orgasmic experience. Despite Stern's claim that this was the best sex he'd ever had, just imagine how much better it might have been for him if instead of a lifeless doll, with no intelligence, no conversation, and no sparkle, he had pleasured himself with a fembot who told him how much she loved him and what a wonderful lover he was, who caressed him and employed her other sensual capabilities to heighten his enjoyment of the encounter. But more about sex dolls and the technology of the orgasm in chapter 7. For now we return to consider the other reasons, apart from the pure physical pleasure of orgasm, why we enjoy sex.

I feel certain that some readers, despite having digested the evidence and arguments in the earlier chapters, do not yet believe that rational humans will develop emotional attachments for robots by midcentury, let alone be falling in love with them. But those readers will surely admit that a fembot or malebot who not only gives great orgasms but also relieves one's sexual tensions, provides new sexual experiences, leads a path away from boredom, and reduces stress could make an outstanding lover. So even in the absence of a strong emotional attachment from the human side, there will be ample motivation for a significant proportion of the population to desire sex with their robots. For example, the 60 percent of college men in James Carroll's survey who did *not* respond that for them emotional involvement was always or most of the time a prerequisite for having sex—they will be likely customers. Similarly will the 51 percent of men in DeLamater's study who mentioned lust and physical pleasure as their main motivations for engaging in their most recent sexual encounter.

Those readers who *do* already accept the concept of humans falling in love with robots can add to the list of benefits in the previous paragraph more of those in Tables 2 to 4—the ones derived from the emotional attachment that loving owners will feel for their electronic sex partners: the expression by the robot's owner of their closeness and/or love for the fembot or malebot, the giving of pleasure to the robot partner, obtaining reassurance about the robot's virtual love for its owner, the enhancement of one's self-esteem on being praised by the robot for one's lovemaking skills, and satisfying the robot partner's stated desire for sex.

Even for those motivations found in the lower reaches of the survey statistics—those relating to the human's power and domination of the sex partner, giving the human a feeling of sexual conquest, and to drink or drugs as the stimulus for having sex—there will certainly be some occasions when these motivations provide sufficient impetus for having sex with a robot, even for people who are not impelled by any of the more powerful motivations.

This leaves only one motivation from Tables 2, 3, and 4 that cannot be applied to human-robot sexual activity—the desire to procreate with a robot.*

Thus far in this chapter, we have explored the main reasons people decide to have sex with people and why people in the decades to come will decide to have sex with robots, but this discussion has somewhat ignored an important catalytic effect that increases the likelihood that a sexual encounter will take place—the sex appeal of the prospective sex object. In this sense the bare statistics expressed by the survey respondents present only part of the picture. The other part, the seductive part, is less obviously a reason to a participant in a psychological survey, largely because of the natural inclination to rationalize

---

*The idea of human-robot procreation is not as ludicrous as it first appears. In *Robots Unlimited*, I describe some of the self-reproducing robots that have already been created by scientists at Brandeis University, robots that can design and build other robots, including exact replicas of themselves. My description includes the explanation that in future decades a robot will have the capacity to find certain characteristics in its human owner appealing and to design those characteristics into the next robot that it builds.

when answering a questionnaire rather than to admit to being influenced by factors that are not directly related to sexual decision making.

These "other factors"—the behavioral ones, the seductive ones—have been investigated by David Bass at the University of Michigan, who ranked various "male acts" and "female acts" according to how effective they were assessed to be in leading the person's date to the bedroom. In Bass's list of the "twenty most effective male acts," there are nine that could apply to robots, including the top three.

> 1st—He displayed a good sense of humor.
> 2nd—He was sympathetic to her troubles.
> 3rd—He showed good manners.
> 6th—He offered to help her.
> 14th—He smiled a lot at women.
> 15th—He gave encouraging glances to girls.
> 18th—He touched her.
> 19th—He made up jokes to make women laugh.
> 20th—He expressed strong opinions.

In any of these assessments, gleaned mostly from the comments of close friends of the women who admitted being influenced by these "acts," we can replace "he" (the woman's sexual partner) by "it" (a robot) with no loss of validity. Already there are computer programs that can make up new jokes.* Most of them are not wonderful jokes, but some are clever enough to get a smile or a laugh. As the software technologies of joke making develop, so robots will come to appreciate jokes made by their conversation partners—in other words, to have a sense of humor. These robots will ably perform the first and nineteenth acts on the above list.

Being sympathetic, well mannered, and helpful and being able to express strong opinions during conversation are attributes that come from a combination of empathy and conversational skills—nothing here is beyond the bounds of reasonable expectation for the artificially intelli-

*For example, Kim Binsted's program JAPE, which creates puns.

gent robots of 2050. As for smiling and giving encouraging glances, David Hanson's moving head* can already accomplish both. And touching, of course, is just about the easiest thing to design into a robot.

The "twenty most effective female acts" listed by Bass strongly mirrored the male list, omitting only two acts—touching and expressing strong opinions—and including one act not on the male list—telling him things he wanted to hear (another straightforward task, once robots have reached the necessary level in their conversational skills).

]]]]] Sex as a Result of Transference

Transference is a psychological phenomenon, typically described as a subconscious redirection of feelings from one person to another. Whereas attachment is a transference of positive feelings that develop first and specifically for a baby's primary carer, usually its mother, and later in life to objects important to that baby/child/adult, and possibly to other people in the form of romantic love, transference is a redirection of feelings, positive or negative, that were first associated with a significant person in the subject's life, not necessarily its primary carer, and are later transferred toward some other person. As an example, one might have negative feelings toward somebody whose manners, voice, or appearance resembles that of an abusive parent, a sadistic teacher, or physical education instructor, a bully or a tease, or a loathed ex-spouse. Examples of positive feelings also abound and might be more closely related to sexuality: a dazzling girl who sat in front of a boy in their high-school algebra class, inspiring his sexual fantasies, or a sexy teacher whose slit skirt and abundant cleavage were similarly inspirational. Transference was first described by Sigmund Freud, who recognized that the models we create in our minds during our formative years as to how people behave stay with us and affect our choices, experiences, and relationships into adulthood.

John Suler, in his fascinating article "Mom, Dad, Computer (Transference Reactions to Computers)," published on the Psychology

*See page 163.

of Cyberspace Web site, explains how the phenomenon of transference extends to relationships with computers:

> These models also shape how people select and experience things in their lives that are NOT human, but so closely touch our needs and emotions that we want to imbue them with human characteristics. We humans can't help but anthropomorphize the elements in the world around us. It's in our blood. We use our internal models to humanize and shape our experience of cars, houses, pets, careers, the weather . . . and COMPUTERS.
>
> Yes, computers can be a prime target for transference, because they may be perceived as humanlike. They are complex machines that almost seem to "think" like humans think. In fact, some people say they WILL someday be able to "think" like us. Unlike TV, movies, or books, they are highly interactive. We ask them to do something and they do it—at least they usually do (like humans, they sometimes disobey and surprise us). With the new generation of highly visual, auditory, and customisable operating systems and software applications, we also have a machine that can be tailored to reflect what we expect in a companion. The science-fiction fascination with robots and androids is the culmination of this perception of machines as being almost like one of us.
>
> What makes computers especially enticing targets for transference is that they are VAGUELY human and PROGRAMMABLE to be whatever we make them out to be. Psychoanalysts discovered that if they remain relatively ambiguous and neutral in how they behaved with their clients, the clients would begin to shape their perceptions of the analyst according to their internal models from childhood. When faced with an indistinct, seemingly malleable "other," we instinctively fall back on our familiar mental theories about relationships and use those theories to shape how we think, feel, and react to this new, somewhat unclear relationship. This whole process is often unconscious. We are so used to these old templates that they automatically start to mould our perceptions and actions without our really thinking about it.[1]

LOVE AND SEX WITH ROBOTS

Suler's article continues with a discussion of the various ways in which transference can apply to computers—how we might subconsciously experience our computer as being like our mother or father or a sibling. One of these ways is examined in the context of Freudian psychology, in terms of sexual desires and fantasies experienced in relation to one's parents, a subject explored more fully by Norman Holland in another article from the Psychology of Cyberspace Web site. Quoting Joseph Weizenbaum's reaction in 1976 to the way that people anthropomorphized and became deeply involved with his programs ELIZA and DOCTOR, Holland points out that people form bonds with computers more quickly than with other objects:

> The computer just makes this process faster and more drastic, because it exhibits "intelligent" behavior like another human.
>
> In sum, then, we have some fantasies about the computer as a thing: phallic fantasies of power and oral fantasies of engulfing pleasure. We also have these more remarkable fantasies that the computer is something more than a thing, something between person and thing. We have a quasi-human relationship with the machine as helpmate, as true friend, as permissive parent, as sex object, and as sex partner.[2]

Suler describes Holland's view of the computer as being "seen as seductive, as a sex object, a satisfier of desire, as a symbol of sexual power and prowess." Thus the concept of transference to computers has rapidly become a discussion of computers as sex objects, complementing our analysis of the reasons for having sex (with people) and the inference that the same reasons mostly apply to having sex with robots. This analysis demonstrates that humans have the capacity for desiring sex with a robot, the capacity to be seduced by them, while Suler and Holland show that computers, and hence robots, have the capacity to entice us, to seduce us.

> > > > > > > > > >

6 | Why People Pay for Sex

I pay for sex because it is the only way I can get sex. I am not ashamed of
paying for sex. I pay for food. I pay for clothing. I pay for shelter. Why
should I not also pay for sex? Paying for sex does not diminish the pleasure
I derive from it. I am proud that I can afford to pay for as much sex as I
need. Indeed, I sometimes pay more than asked or expected.

—Hugh Loebner[1]

The idea of satisfactory sex being available whenever it is desired
is an extremely appealing one to enormous numbers of men and women,
but for various reasons many people do not enjoy this level of sexual
availability within the confines of a relationship or marriage. This might
be because they are not in a relationship at all, or because they are often
away from their partner because of business or other travel commitments, or because their relationship partner does not enjoy sex (with
them) as much as they do (with their partner) or is not a good sex partner in some other respect. Whatever the reason, this void in their life
has a simple remedy that for thousands of years has been adopted by a
small but significant proportion of the sexually active population. The
remedy is to pay for it, and the prevalence with which paying for sex has
existed for so long has enabled the world's oldest profession to survive
and even at times to thrive. Soliciting the services of prostitutes provides a relatively easy cure for sexual frustration, and there is no evidence to suggest that those with the necessary financial means need
ever go without sex through a lack of supply. Far from it.

In this chapter we examine the reasons people pay for sex with
the women and men who ply this particular trade. From the perspective of sex with robots, what is interesting about the most frequently
proffered reasons is that they indicate desires, and not only the desires
for the sex acts themselves, that could be satisfied by a sophisticated

robot just as well as by a human prostitute. This being the case, it seems inevitable that just as humans desirous of sex but lacking sufficient opportunity will pay a professional for it, so there will come a time—and that time is almost with us—when people will be paying for sex with robots, either by buying the robot for regular use at home or by renting one by the hour or the day. The enjoyment and benefits derived by their owners or renters from the sex they experience with robots can reasonably be expected to bring as much overall satisfaction as those same people enjoy as the clients of (human) prostitutes.

We commence our discussion of the parallels between paying human prostitutes and purchasing robot sex with a comparison between the reasons men pay for sex and the reasons women do so. The purpose of this comparison, in part, is to support the argument that women will become just as eager as men to seek sexual satisfaction from robots.

]]]]] Men Paying Women

Obtaining accurate estimates of the percentage of the population that visits prostitutes is fraught with difficulties, largely due to the stigmatizing view of prostitution and its clients that has long been held by so many people. As a result, serious attempts at quantifying the use made of the services of prostitutes did not begin until the middle of the twentieth century, when Alfred Kinsey estimated that 69 percent of the white male population of America had been to a prostitute at least once in their lives. If this figure seems high to some readers, it should be considered alongside the historic study by Timothy Gilfoyle, who estimated that 10 to 25 percent of *all* young women in New York City in the nineteenth century were prostitutes, either temporarily or on a long-term basis, making prostitution for much of that period the second-largest business in terms of the revenue generated (the first being tailoring). It should also be noted that Kinsey's figure of 69 percent was for men who have had sex with a prostitute *at least* once in their lives, and most of these men had only one or two such experiences. Kinsey's estimate of the proportion of American men whose sole sexual outlet was prostitutes was very much lower, between 3.5 and 4 percent. Other, more recent estimates by different researchers of the numbers of men in the United States who had had sex for money have ranged from 16 percent (estimated in 1994), to 18 percent of those aged eighteen to fifty-nine (in 1998), to 20 percent (in 1993).

The accuracy of all these figures must be viewed with some doubt because of the known discrepancies between what johns are willing to admit in an interview survey or when filling in a pencil-and-paper questionnaire and the figures ascertained by other means that are known to be more reliable. This phenomenon of underreporting by johns was examined in a study published in 2000 by the U.S. Academy of Sciences, entitled *Prostitution and the Sex Discrepancy in Reported Number of Sexual Partners*. In their report, Devon Brewer and his colleagues found that when responding via a computer-assisted interviewing process, which is generally believed to promote accurate reporting, the johns' answers relating to contacts with prostitutes were almost four times higher than when responding to human interviewing or pencil-and-paper questionnaires. It seems most unlikely (not to say impossible) that this four-to-one ratio persists across the whole spectrum of quantitative research on prostitution, but what *is* clear is that figures such as 16 to 20 percent in the United States should certainly be regarded as understatements.

In many parts of Europe, it has long been common practice for young men to receive their sexual initiation from a prostitute, though the numbers might be declining due to changing moral values that no longer place such stringent constraints on the sexual behavior of young unmarried women. A study carried out in Lisbon found that 25 percent of men in their sample of 200 had lost their virginity to a prostitute, while in a study in France some thirty years earlier, 47 percent of men who were practicing Catholics similarly had their first sexual experience in this way. Given that these particular statistics do not include those men whose *first* visit to a prostitute came when they were no longer virgins, it would appear that the overall figures for men who have had sex with a prostitute is significantly higher in these strongly Catholic countries than one might otherwise expect.

Estimates from other developed countries vary considerably. From a national study on sexual attitudes and lifestyles carried out in Britain during the early 1990s, out of 19,000 households surveyed, only 1.8 percent of men responded that they had paid for sex during the previous five years. A second survey, published seven years later, noted an increase to 4.3 percent, but the authors questioned whether this was a genuine increase in numbers or whether it was because those surveyed for the later report were more willing to admit to their peccadilloes. Other studies include a 1991 national telephone survey in Switzerland, which estimated that 12 percent of men between seventeen and thirty years of age had visited prostitutes, and a study at about the same time by Cecelie Hoigard and Liv Finstad in Norway, that estimated the figure to be 13 percent. The difficulty in obtaining accurate estimates, even nowadays when people are more willing than in the past to discuss their sexual habits, is shown by the results of a survey in Holland: Only 3 percent of heterosexual men aged eighteen to fifty were willing to admit to having paid for sex in the previous year, whereas calculations based on the estimated number of prostitutes and the average number of clients that they serve per day put the figure at 16 percent. On other continents estimates are significantly higher, particularly in developing countries such as Thailand and the Philippines. In 1994 it was estimated that each day more than 450,000 Thai men visited prostitutes, while "prostitution, as an integral component of the tourist industry, is an important source of foreign exchange for the Philippine Government."[2]

These somewhat diverse statistics confirm that even though there is a wide variation in the percentages between countries, a huge number of men employ the services of prostitutes.

]]]]] Women Paying Men

Some writers have thought that if buying sex is a benefit for men, it must also be a potential benefit for women, one they should be encouraged to seek out. Ericsson,* for example, argues that under the present unequal circumstances of sex work, "some benefit is withheld from or denied women that is not withheld from or denied men. The best way to deal with this inequality would not be an attempt to stamp out the institution but an attempt to modify it, by making the benefit in question available to both sexes."

—Christine Overall[3]

It has always been the case that the number of male clients of female prostitutes far outweighs the number of women who pay men for sexual services. The principal reason for this, pointed out by Kingsley Davis in his 1937 article "The Sociology of Prostitution," has been economic—the number of women who earned enough (or had jobs at all) to allow them to pay for sexual services has been considerably below the corresponding number for men. Nevertheless, the practice has existed since at least the late nineteenth century. In *The Sexual Life of Our Time,* Iwan Bloch refers to an anonymously written 1848 book, *Prostitution in Berlin and Its Victims,*† which

> contains an appendix on "prostituted men" (p. 207), who, however, are not homosexual prostitutes, but, according to the writer's own definition, "men who make it their profession to serve for payment voluptuous women by the gratification of the latter's unnatural passions." This species still exists to the present day [i.e., 1909], but there is no particular name for the type. (In the seventies [the 1870s], in Vienna, men who could be hired to perform coitus were known locally as "stallions"—German *Hengste.*)

In the pleasure-seeking boom of the post-prohibition United States, it was inevitable that men-for-hire-for-ladies would become a growth area within the world's oldest profession. Ted Peckham quickly became famous in New York society during the mid-1930s for being able to supply presentable men who would satisfy the desires of his

*Lars O. Ericsson, in a 1980 article in the journal *Ethics 90.*
†*Die Prostitution in Berlin und ihre Opfer.*

largely wealthy female clientele, on a strictly pay-as-you-go basis with a charge for overtime after midnight. For four years his agency, Guide Escort Services, was a booming success and very much in the public eye, even opening for business in Europe, but eventually the law turned against Peckham in the form of a writ accusing him of running an employment bureau without a license, a legal ploy designed to get around the problem that the authorities doubted whether any charges filed against him relating to prostitution could be made to stick. Peckham was prosecuted by the forceful gangbuster and district attorney Thomas E. Dewey, who later became governor of New York and was a Republican candidate for the presidency in two elections.* Peckham was found guilty by the judge (no jury) and was fined $250, with an additional sentence of three months in the workhouse suspended during his "good behavior" and "upon the condition that he not conduct this agency unless and until he has obtained from the proper authorities of the city of New York a license to do so." Peckham duly gave up the escort business and became a writer.[†]

Peckham may have been the exception rather than the rule during the 1930s. His notoriety did little to dent the assumption by most people that women have no need or wish to pay for sexual services, a view that prevailed at least until the advent of the boy-toy fashion in the early 1990s. This fashion, and the changing behavior patterns that accompanied and followed it, were all part of the new era of feminism, which encouraged women to assert their equal right to full and satisfactory sex lives. Enter Joel Ryan, a twenty-first-century version of Peckham, who runs a successful "escort" business called Heaven on Earth in Melbourne, Australia, catering to both male and female heterosexual clients (women make up approximately 40 percent of his client list). Brothels and escort services may legally ply their trade throughout much of Australia, as a result of which Ryan and his service

---

*In 1944 and 1948.
[†]The dust jacket of his memoires, *Gentlemen for Rent*, proclaims that his service in New York was launched with the blessing of a host of celebrities, including Lucius Beebe, Maury Paul, Walter Winchell, Danton Walker, and Louis Sobol, and that in London he was "sponsored by the Duke of Kent."

have become something of a curiosity item in the media, including the subjects of a television film, *What Sort of Gentleman Are You After?* by the British documentary maker Jane Treays.*

While Joel Ryan serves both men and women clients, a new brothel service announced in Valencia, Spain, was set up in 2006 *by* a woman exclusively *for* women. In Spain, as in most countries, visiting prostitutes is traditionally seen as *una cosa de hombres* (a men's thing), with an estimated 25 percent of men having indulged, according to a survey by the Institute for National Statistics. And while the medical publisher Mundo Médico's figure for Spanish women who have paid a gigolo is very much smaller at 2 percent, it is nevertheless higher than many people would expect, especially for a strongly Catholic country.

"Charming Barbara," the madam who opened and runs this particular Valencia brothel, was herself a sex worker for eight years. Then she set up an agency for female clients, offering male escorts, but soon decided to start a permanent luxury brothel. Barbara has had no shortage of men who want to work for her and is very clear about what her mostly professional executive clients want for their money, which can reach about €1,200 ($1,500) for a whole-night session: "I don't want muscle men. Above all they must have good conversation."

The advent of the Internet has greatly facilitated prostitution, by making it possible to advertise, almost free of charge, to a huge potential client base. This freedom is being exploited by increasing numbers of men who advertise their sexual services to women in language that often leaves little to the imagination in terms of the advertiser's claimed sexual prowess and size. In 2005, Isabel Kessler, at Middlesex University, investigated this growing trend. She found that between 150 and 200 male escorts offered their services to women in London via their own Web sites, which could be viewed free of charge. Kessler did not investigate the number of men advertising on so-called membership home pages, for which access is available only upon payment of a fee, so the figure of 150 to 200 can safely be assumed to be an underestimate. Typical of the

---

*Shown as part of the BBC's *Under the Sun* series on December 15, 1997, watched by an estimated 4.8 million viewers.

charges quoted by these sites at that time were £100 ($180) per hour and around £450 ($800) for an overnight session.

The comparatively recent growth in the heterosexual male prostitute business in the United Kingdom is almost nothing in comparison with a phenomenal surge in demand from financially well-off women in Thailand, noted in 2002 by Zenitha Prince in her article "Thai Female Elite Demand Black Gigolos," which appeared in the independent newspaper of Morgan State University:*

The long-perpetrated image of the black man as a sexual toy continues to flourish as the niche market for black male prostitutes in Thailand booms.

Escort services are now importing hundreds of prospective black gigolos from Jamaica and Africa into the Asian country to satisfy the surge in the demand for these services among Thai female elite.

A research project, recently completed by Associate Professor of Sociology Nither Tinnakul, from Bangkok's Chulalongkorn University, puts the number of male prostitutes in Thailand at a staggering 30,000, triple the estimated amount of just two years ago. . . .

"I think the women want some equal rights you know [revenge against philandering husbands], some kind of freedom. She needs something," Tinnakul said.

Apparently, this is a need that these black foreign prostitutes or "forungs" have aptly satisfied. The report further stated that Thai women are paying upwards of 10,000 baht (243 dollars) per night for the servicers, who are "fiercer," more "thrilling" in bed than their Thai peers and "well built."

A different form of prostitution for women clients has also been rising steadily in popularity in recent years—sex tourism, which is also referred to, with the benefit of a fair dose of delusion, as "romance

*November 8, 2002.

tourism." Sex tourism has of course long been popular with many men, who travel to Thailand, the Philippines, Bali, and elsewhere in the knowledge that the price of sex in their chosen destination comes very cheaply. For most of these men, the transaction is simple prostitution, sometimes for a single brief encounter and sometimes for longer, perhaps for most or all of their vacation if they meet a girl who gives them a really good time. For a few others, it is a means of finding a satisfying wife to take home.

For women both the nature of the transaction in sex tourism and the treatment they are seeking differ from those of male sex tourists. Instead of a cash transaction that is overtly money for sex, often paid in advance, the payment comes in ways that the woman can rationalize as a gift, helping out the beach boy or the tourist guide and his family. "Most beach boys enter into sexual relationships with as many tourist women as they possibly can, and most of these relationships result in some form of material or economic benefit for the man. Some beach boys and hotel or bar workers engage in explicit sex for cash exchanges with male tourists, female tourists and/or tourist couples, but on the whole, the economic element of their sexual relationships with tourist women is less formally arranged."[4] These men play the game of pretending to be genuinely attracted to the women, of falling in love with them and wanting to marry them.* In turn the woman plays the game of enjoying being pampered and often deludes herself into believing that the man loves her and that she loves him. She buys him meals, buys him presents, and gives him money for a "sick relative" or on some other pretext, often repeating the cash gift after she returns home from her vacation. The whole process is described by Nigel Bowen in his article "Sugar Mamas":

> It is not sex for sale; it is love for sale. These guys get girls by courting them, charming them, wooing them. Women are attracted to

*Sometimes there is indeed a desire to marry, as that will often provide the man with a First World passport and therefore an exit from his Third World poverty, and sometimes the woman will return to the same vacation spot and to the same man, eventually sending him a plane ticket so that she can import him to her own country.

the romance of it. It is a fantasy to meet an exotic stranger on the street who seems to have fallen in love with you at first sight. Balinese men target women's hearts: they're sensitive, sweet, flattering and funny. And they're also very clever about going for the Achilles heel. If a girl is fat, they'll tell her she has a beautiful body.

Prue (not her real name) is an exuberant fifty-four-year-old widow with a healthy bank balance and an even healthier libido. Three times a year, she locks up her home in one of Sydney's more respectable suburbs and flies to Bali for the sole purpose of spending a week being sexually pampered by a teenager. "The Balinese say they love you, and of course I want to hear that, but at the end of the day, it is a business deal. At my age, money is their sole focus. I pay for the accommodation, meals, excursions and buy them gifts. At the end of the holiday, I slip several thousand dollars—enough to support their family for six months—into an envelope and leave it on the table for them."[5]

From the little published research that exists on the prevalence of women sex tourists, it appears that in some tourist destinations at least the practice is rapidly becoming commonplace. Jacqueline Sánchez Taylor surveyed 240 women who were on vacation alone in the Dominican Republic and Jamaica, asking them to complete a questionnaire for a study on tourism and sexual health. "A questionnaire was constructed which was designed to yield some basic data on tourist women who had sexual contact with local men, including their nationality, age, occupation and racialized identity; their perceptions of the 'sexual culture' of the host country; how often they had traveled to that country and other known sex tourist destinations; how many different local sexual partners they had and whether they perceived these relationships as 'real love,' 'holiday romances' or 'purely physical'; whether or not they gave money or made other gifts to their local sexual partners; whether or not they took safe-sex precautions."[6] When responding to the questions about how they perceived their relationships with their local lovers, 39 percent of the women described it as a holiday romance, 22 percent as real love, and only 3 percent as purely physical. (A further 12 percent said it was both physical and a holiday romance.) Taylor found that part of the self-delusion process is due to "racist ideas about black men being hypersexual and unable to control their sexuality," which enables the women "to explain to themselves why such young and desirable men would be eager for sex with older, and/or often overweight women, without having to think that their partners were interested in them only for economic reasons . . . Only women who had entered into a series of brief sexual encounters began to acknowledge that 'it's all about money.'"

Almost one-third of those who completed Taylor's questionnaire admitted to engaging in one or more sexual relationships with local men during the course of their holiday. These women ranged in age from girls in their late teens to women in their sixties, the most likely to indulge being those in their thirties to forties. About a quarter of the women surveyed said that they had been offered sex for money by local men, but not one of these woman admitted to have taken up the offer, so those who did engage sexually with the locals clearly did not accept that there was a commercial element to the relationship. This is despite the fact that 57 percent of the women who did take local sex partners acknowledged that they gave their lovers "help" in the form of cash, gifts, and/or meals. Taylor recognizes that because of underreporting, this figure "is unlikely to accurately describe the true level of economic benefits transferred to local men by these women."

Taylor also found that these women differ in terms of the type of sexual encounter they are after and the manner in which they rationalize these encounters. "Some are eager to find a man as soon as they get off the plane and enter into multiple, brief, and instrumental relationships; others want to be romanced and sweet-talked by one or perhaps two men during their holiday."[7]

]]]]] Why Men Pay Women for Sex

Several reasons have been identified as to why men pay women for sex—what the men want or expect from these sexual encounters. While the reasons vary somewhat from one country to another, there is one common underlying emotional need that appears to be extremely widespread. It is the need for mutuality, the self-delusional feeling that the prostitute is a true *partner* in a relationship, however brief. This "myth of mutuality," as Elizabeth Plumridge calls it, posits the typical prostitute as caring about the client and enjoying her intimacy with him. For the johns interviewed by Plumridge for her study, all of whom were clients at a New Zealand massage parlor, "pleasure rested on two postulates; on the one hand a complex of notions that revolved around relaxation from constraints and obligation, and on the other, a set of interpretations that relates to mutuality."[8] Plumridge found that these men wanted the myth of social warmth to be sustained from the moment they entered the so-called massage parlor and would complain "if the surface social pleasantries were torn away and the naked imperatives of sexual exchange for money revealed as the true purpose of the warm reception." The johns in her study did not all claim that the prostitutes they visited loved them, but all of these men did ascribe some level of emotionality to the encounters, describing their visits in terms such as "very nice to be pampered, just the feel of it and the warmth."

This desire for reciprocity perhaps explains certain trends in the U.S. sex industry in recent years, away from brief gratification for the man and toward a warmer, more sociable environment for the sexual encounter, as explained by Elizabeth Bernstein:

> Those [johns] who frequented indoor venues enjoyed the benefit of an arrangement that was structured to more effectively provide them with the semblance of genuine erotic connection. For example, interactions with escorts as opposed to streetwalkers are typically more sustained (averaging an hour as opposed to 15 minutes), more likely to occur in comfortable settings (an apartment or hotel room, rather than a car), and more likely to include conversation as well as a diversity of sexual activities (vaginal intercourse, bodily caresses, genital touching, and cunnilingus, rather than simple fellatio). The fact that street prostitution now constitutes a marginal and declining sector of the sex trade means a transaction that has been associated with quick, impersonal sex-

ual release is increasingly being superseded by one which is con-figured to encourage the fantasy of sensuous reciprocity. . . .

In recent years, one of the most sought after features in the prostitution encounter has become the "Girlfriend Experience," or *GFE*. In contrast to commercial transactions premised upon the straightforward exchange of money for orgasm, clients describe the GFE as proceeding "much more like a nonpaid encounter between two lovers," with the possibility of unhurried foreplay, reciprocal cuddling, and passionate kisses.[9]

Several of those interviewed by Plumridge claimed that their paid sex partners were of greater emotional importance to them than their relationships with their wives. While this speaks volumes about the states of these men's marriages, the fact that they genuinely believed it, or at least deluded themselves into believing it, demonstrates just how easy it is for someone who *wants* a particular person to care about him to succumb to the myth that that person does indeed care. Plumridge summed this up by explaining that the men "all wanted a responsive embodied woman to have sex with. This they secured by ascribing desires, responses and sexuality to prostitute women. They did not know the true 'selves' of these women, but constructed them strategi-cally in a way that forwarded their own pleasures." Another New Zealand study on why men visit prostitutes found further support for the companionship, the myth of mutuality, and the lack of complica-tions as prime reasons for paying for sex. There was also an emphasis by some of the johns in this particular study on the inability of their wives to satisfy them sexually.

Some fifteen years prior to Plumridge's research, Harold Holzman and Sharon Pines had examined the motivations of a sample of men aged from twenty-seven to fifty-two almost half of whom were married, for their paper "Buying Sex: The Phenomenology of Being a John." In common with Plumridge, they found that the men's desire for sex was coupled with a desire for companionship, hence "in every encounter discussed, the individual paying for sex engaged in social, courting behaviors that were often flavored with varying degrees of romance."[10]

Holzman and Pines found that "there existed a belief that by being pleasant or even quite amorous they could subtly seduce the prostitute into allowing their created illusion to play itself out . . . Clearly, a great deal of energy is invested in the maintenance of the illusion." Roger Kernsmith found the same need for a social bond in those johns whose postings on the Internet newsgroup ASP (alt.sex.prostitution) he studied for his survey: "The theme that clients hired prostitutes as much for the sense of social closeness and acceptance as for the physical stimuli associated with the performed sex acts was found in every element of the ASP data."[11] Further reinforcement of this theme comes from an even earlier study by Charles Winick, who interviewed 732 clients in five major American cities and concluded that the "emotional meanings and overtones of a client's visit to a prostitute are more important to the client than the desire for sex."[12]

The importance of companionship for the client as a benefit of the transaction is fully acknowledged in the teaching at the Hanky Panky School in Amsterdam, which was opened in 2003 by Elene Vis, the former madam of an escort agency. In the Netherlands, brothels are legal and prostitutes pay tax on their earnings. The country is renowned for its tolerant attitude toward commercial sex, and the red-light district in Amsterdam is a well-known tourist attraction where the women display themselves in shop windows.

Vis prides herself on teaching her students to perfect their skills and boost their sales by giving "more than sex." "Of course I teach sex techniques to the students, but with a client that only takes ten minutes and does not satisfy the customer or the escort. . . . It is about attention, listening, tenderness and positive energy, and those things can be bought."*[13]

*And for those men who wish to take up heterosexual prostitution as a career, and to learn the tricks of the trade, there is an online gigolo school called Gigolo International that will doubtless teach you everything you need to know. The site advertises membership for $49.95, with the news that "Modern upscale (working) women are always busy and have become more emancipated regarding paid (erotic) company. This could be a dinner-date, a short business-trip or even a fully paid vacation! These women need a man to share quality time without troubles afterwards and they will gladly pay for the right services. Did you ever dream of becoming

Vis claims that after a half day of classes, the prostitutes emerge with the power to change "ten disappointing minutes" into an exchange of positive energy. "Escort is accompanying someone. If the man feels pampered, he will be willing to pay for more than just the sex. In return this boosts the girl's self-esteem." For €450 ($490) students can take classes in "Presentation," "Adding On Hours," and "Entertaining."*

The illusion in the minds of the johns, this myth of mutuality, is something that will be even more believable when the pampering and the imitation affection emanate from a sexual robot rather than a human prostitute. One reason for this is that in the case of the prostitute the john pays for every encounter and is therefore reminded repeatedly of the connection between the sexual experience and money, whereas when he's purchased a robot, this connection in its owner's mind will quickly dissipate forever.[†] More obvious reasons why the robot experience will be more appealing than visiting a prostitute include the utterly convincing manner in which robots will express affection and other emotions, simply because their emotions will be programmed into them, to be part of them, instead of being make-believe affections acted out by a prostitute with little genuine enthusiasm for the need to convince.

Gigolo? This could be your first step into a whole new lifestyle! The members-section contains: Gigolo's tricks of the trade; How to become an independent Gigolo; Independent Gigolo promotional tools; The Ultimate Gigolo; Exclusive discount for members; Increase your female-contact skills; New tips and updates every month; Earn additional TOP$$ with our unique referral-program."

*The Hanky Panky School inspired a delightful parody of the same idea in India, in a March 31, 2005, article by Sidhi Chadh in the Hindustan Times, entitled "Now Learn Prostitution in School." It commenced, "A Diploma in Sex Trade? That will be among the several qualifications on offer when a government-sponsored school for prostitutes opens in the capital on Friday. The move to encourage sex workers who are fully trained in their craft comes just days after the U.S. threatened to impose sanctions unless the administration did something to regulate the flesh trade in the country." The article also explains that "the girls will learn everything from seduction to handling finances. Besides giving the girls useful tips about sex, we will also tell them how to seduce clients and extract maximum money."

†Of course, there might be robot prostitutes, for johns who lack the resources or the inclination to purchase a sexual robot for use at home, and if there are robot prostitutes, then there will also be robot brothels, staffed by robots for the benefit of humans.

Motivation: Variety

Neil McKeganey and Marina Barnard studied the reasons most often mentioned by johns as being important in the decision to employ the services of prostitutes. One reason they highlighted is variety—the opportunity to have sex with a range of different women. Plumridge's research confirmed this finding, quoting as an example one john who explained his motivation as "someone different someone new," and another responding that "What actually turns me on is a bit of variety around me."

A robot will be able to provide endless variety in terms of its conversation, its voice, its knowledge and its virtual interests, its personality, and just about every other aspect of its being. All will be changeable on demand. Even a robot's physical characteristics could be changeable, thanks to clever mechanical design and replacement parts. And all aspects of a robot's sexuality will similarly be changeable according to its owner's wishes. It is hardly practical for a john to go searching the streets of a red-light district or to a brothel expecting to be able to find a woman looking like Marilyn Monroe (or whoever his lust desires), with the brainpower and knowledge of a university professor and the conversational style of a party-loving teenager. But with a robot at home, he need search no further—all these characteristics and more will be selectable at the time of purchase. So the man who wants variety in his sexual partners will be able to find it, wherever he wishes, and far more easily than when looking for a prostitute to match his desires.

While variety in the appearance, personality, and attitude of prostitutes is one major reason men pay for sex, variety in the sexual experience itself is, for many johns, another important factor, often *the* most important. "One of the main reasons clients pursue encounters with prostitutes is that they are interested in sexual practices to which they do not have access, either because they have no regular partners or because their partners are unable or unwilling to accommodate their desires."[14]

Many men are in relationships in which their wife or partner's sexual tastes do not accept oral sex or some other sexual practice in which the man would like to indulge, so paying for the service provides an easy way out of his problem. The extent to which oral sex is an important reason for male clients to visit female prostitutes was studied by Martin Monto at Oregon's University of Portland and described in his 2001 paper "Prostitution and Fellatio" in the *Journal of Sex Research*. Monto gathered questionnaires from more than 1,200 men who had been arrested while attempting to hire female prostitutes in Las Vegas, Portland, and San Francisco and who were participating in a program known as "johns school," designed to discourage them from reoffending.* The results of Monto's survey indicate that having a prostitute perform oral sex was even more prevalent for most johns than vaginal sex, with 81 percent of those surveyed having experienced oral sex given by a prostitute, when the figure for vaginal sex was only 55 percent. However, when asked how they would rate various sexual activities, 76 percent of the johns in Monto's survey described vaginal intercourse as very appealing, while the figure for "having a partner perform oral sex on you" was lower, at 65 percent. Monto suggests various possible reasons for this discrepancy. One reason is that "it is much easier and more convenient to engage in oral sex than vaginal intercourse in a car or alley, where many of these episodes occur"; another is that prostitutes might "prefer oral to vaginal sex for a variety of reasons"; while others include clients' beliefs that there is a lower risk of AIDS from oral sex than there is from vaginal sex and less need to wear a condom.[15,16]

Comparing Monto's figures with findings from the National Health and Social Life Survey† revealed that in the U.S. male population as a whole, a significantly lower proportion of men, 45 percent, found receiving oral sex very appealing. This comparison, between johns and the male population as a whole, indicates that a significant number of men seek encounters with prostitutes because of a desire for oral sex, providing a convincing example of the wish for sexual variety as a prime motivator for many johns. This motivation has also been confirmed by various other researchers.

*Attending these programs wipes out the record of the participant's arrest, thereby ensuring that almost all those who are arrested for attempting to hire prostitutes in these three jurisdictions take up the offer of attending a johns school.
†A survey conducted among 3,422 respondents between February and October 1992 across all fifty states of the United States.

Clearly there will not be much of a problem in designing a mechanically sophisticated robot so that it can perform oral sex. Furthermore, sexual robots will be able not only to satisfy any particular sexual desire expressed by their owners but also to suggest sexual practices that their owners have never previously experienced and to teach their owners to become better lovers for those occasions when they prefer human company to sex with their robot.

Motivation: Lack of Complication and Constraint

> For many clients, one of the chief virtues of commercial sex exchange is the clear and bounded nature of the encounter. . . . What is unique to contemporary client narratives is the explicitly stated *preference* for this type of bounded intimate engagement over other relational forms.
>
> —Elizabeth Bernstein[17]

Alongside variety as a prime reason for visiting prostitutes, the research literature has identified a small group of motivations that might collectively be described as a lack of complications and constraints. Neil McKeganey found that "for some men the appeal of prostitution seemed to lie in a combination of the anonymity, the brevity and the emotionally uninvolved nature of the prostitute contact."[18] Another survey, this one conducted among the clients at two Australian brothels, indicated that 90 percent of the men who participated confirmed that a lack of "complications" was one of their main motivations in paying for sex. From earlier research it appears that the complication most often cited as an obstacle to getting noncommercial sex was the perceived need by men to "play games," pampering and courting a woman in order to achieve their goal, possibly requiring an enormous effort in return for which there is no guarantee of sex. As one of the johns in Holzman and Pines's survey put it, "If I just want to go out and get laid I'm not going to bother going to a bar and buying drinks and dancing with a girl all night because I'm not interested in that . . . you don't want to spend time looking for it where there is always a maybe—maybe yes or maybe no . . . you almost want a written guarantee."[19]

When discussing their wish for sex without constraints, johns present something of a paradox in their attitudes. On the one hand, there is the self-delusion of the myth that a measure of emotional involvement exists in both directions with the prostitute. In contrast there is what Monto describes as "a wish to avoid the responsibilities or emotional attachments of a conventional relationship,"[20] the attitude that "payment of money for sex entitled them to freedom from the requirements normally associated with relationships."[21] Many of the johns who provided data for this research indicated that they regarded sex with their wife or partner as part of a different type of transaction, one in which they were tied down or had other demands placed on them. So instead of playing games, instead of the constriction of obligations imposed within a noncommercial sexual encounter, the johns are attracted by the ease of a paid sexual experience. The limited nature of paid sexual encounters and the lack of any long-term emotional involvement further contribute to the johns' feelings of freedom. The payment of cash is a simple, direct way to guarantee a sexual experience with the minimum of effort.

To avoid any necessity to indulge in games in the pursuit of a sex partner, for the avoidance of what are often perceived by johns as being constraints and complications in more conventional sexual relationships, and in the interest of limiting the nature and duration of any emotional involvement to whatever extent is wanted by its owner, a robot will be the ideal sex partner. You don't have to buy it endless meals or drinks, take it to the movies or on vacation to romantic but expensive destinations. It will expect nothing from you, no long-term (or even short-term) emotional returns, unless you have chosen it to be programmed to do so.

Motivation: Lack of Success with Women

> The basic function of prostitution is to provide a primarily sexual service
> to people who either fail to meet the requirements of the more legitimate
> "market" or who exclude themselves from the larger market because they do
> not feel comfortable in it. The system is very flexible, and no-one is turned away.
>
> —Mary Laner[22]

For a variety of reasons, many men have difficulty in becoming involved in dating or more permanent relationships with women. In some cases this is because the man is ugly, physically deformed, psychologically inadequate, a stranger in another town or a foreign land, or simply lacking in the necessary social skills and/or sexual assurance. Such men, with normal male desires, have a need for sexual intimacy that they cannot satisfy because of their lack of sexual effectiveness—they simply cannot attract a mate, or are afraid to try, or suffer from a combination of both. Their need can, however, be satisfied by a prostitute. By seeking to pay for sex, they reduce the risk of rejection to an absolute minimum, thereby almost guaranteeing themselves sex on a plate. For these men, prostitution is the only sex available, a reason for paying for sex that was indicated by almost 40 percent of the johns in a study by Xantidis and McCabe.

None of these categories will present any problem to robots. Any man lacking in the self-esteem necessary to make sexual overtures will be able to purchase a fembot that is immune to ugliness or a physical deformity in its owner and to its owner's psychological inadequacies.*
She will be available for hire (or travel with her owner) when he is visiting another town or country. And she will cater to the socially inept and the sexually unassured with the same virtual emotions and the same sexual responsiveness as when she is encountered by the handsomest, the most socially adept, and the most sexually confident.

]]]]] Why Women Pay Men for Sex

In contrast to the relatively well researched topic of men paying for sex, there is almost no systematic published research on the reasons why women pay, on what exactly they are seeking from their commercial sexual encounters. Reported anecdotal evidence from a number of media articles on sex tourism promotes the view that when on vacation many women are looking for physical satisfaction from young, toned,

*And given that most men who employ the services of prostitutes do not suffer any loss of self-esteem for doing so, it seems reasonable to assume that most men who use sex robots will similarly not suffer from an undue loss of self-esteem.

native male bodies with large penises. But that is far from being the whole story, as an examination of the extremely sparse evidence testifies. The truth appears to be close to what has been observed from the studies on why men pay women for sex.

One source that does go some little way to explain why women pay for sex is an article published in the UK edition of *Marie Claire* in February 1994: "Why Women Go to Male Prostitutes," the research for which was carried out by an academic at Birmingham University (who was not credited in the magazine*). Ten women were interviewed for this article, of whom nine gave reasons that mirror some of those generally expounded by men.[†]

Three of the women responded very much in line with the "myth of mutuality," wanting social warmth, caring, companionship. Jane,[‡] aged fifty-one, a housewife, commented, "It's not so much the sex I'm looking for, as the feeling that someone is there for me." Jean, a thirty-nine-year-old teacher, endorsed the importance of companionship: "It's not even the sex I want—just the company. Unless you have been through it yourself [husband leaving you for another woman], it's impossible to understand how desperately lonely you get. . . . I did eventually have sex once, but I would be just as happy to pay for the company." And Anne, sixty-four, a housewife, was very much like Prue the sex tourist, in her attitude: "I suppose I think of it as a holiday romance more than anything else. I would never dream of looking for anyone here in England. I certainly didn't feel demeaned by it. I wouldn't expect a man to want to do it with a woman of my age for nothing."

Six of the other women in the *Marie Claire* article also espoused motivations that are among the most prevalent ones expressed by men.

Lack of Complications and Constraints

"The only thing lacking in my life is regular and uncomplicated sex," said Yasmin, forty-four, a charity fund-raiser. And, "The only way I

*Carmen Caldas-Coulthard.
[†]The reason given by the tenth woman was that her husband wanted her to do it (while he watched).
[‡]The names of the women were, of course, changed for publication.

could guarantee sex without involvement was to pay for it," stated Barbara, twenty-nine, a hospital administrator.

Lack of Success with Men

"I have always been overweight and have developed a bit of a complex about it over the years. I've never really had a proper boyfriend—not one that lasted more than a few weeks anyway," said Lucy, thirty-five, a housing officer. "Finding a new man [after a breakup] seemed impossible," according to Nicole, an art director. "My husband hasn't made love to me for ten years," said Irene, thirty-seven, a housewife, describing lack of success with one particular man—the one she wanted. And similarly, "I felt neglected by Colin's lack of interest in me," said Louise, forty-seven, a doctor's receptionist—about her husband.

A small but useful source of additional data on the reasons women pay for sex is the group of clients who consented to be interviewed for the documentary *What Sort of Gentleman Are You After?* Their comments revealed that in addition to simply enjoying "good sex," by and large they are motivated by the same desire for a "lack of complications" that appeals to many johns*:

"A mutual adult consent meet. No bullshit about it."

"It's a completely business transaction."

"The beauty, I think, of paying for it as opposed to picking up somebody is that I feel I didn't need I need to repay the favor. I didn't need to pleasure him. I could just lie there and absorb it all. If I'd wanted him to go down on me for the entire two hours, then I could have said it and he would have done it."

"It is so much easier than having to go out and pick one up, and then if that's all you want, you're left with him there and you can't get rid of him without being extremely rude."

"I tend to go for long periods without sex, basically. Not through

---

*Other reasons given by Ryan's clients in these interviews are: "He's very inventive sexually, and that's why I keep seeing him—it's always fun, it's always something new" (i.e., the variety motive); "I find it quite exciting to pay for it. I find that quite sexy"; and "They've got to have something, a spark, and a big penis as well."

choice, but through not finding anyone I fancy. So I find I end up in relationships for two or three months with complete assholes just to have sex."

]]]]] The Future of Prostitution

With women only recently beginning to swell the number of clients of prostitution, the world's oldest profession is currently thriving and showing every sign of continuing to do so. A study on sexually transmitted infections found that the number of men in Britain who have paid for sex had almost doubled between 1990 and 2000. Of 11,000 men interviewed in 1990, 5.6 percent admitted to having paid for sex. By the year 2000, the figure was more than 9 percent.

I do not believe that this trend will continue forever. Robots will be able to satisfy the myth of mutuality for people of both sexes, to provide variety, to offer sex without complications or constraints, and to meet the needs of those who have no success in finding human sex partners. And for those women who are joining the ranks of today's sex tourists, beautifully toned malebot bodies can be made to order, with whatever vital dimensions are desired.

When sexual robots are available in large numbers, a cold wind is likely to blow through the profession, causing serious unemployment. As long ago as 1983, the *Guardian* reported that New York prostitutes "share some of the fears of other workers—that technology developments may put them completely out of business. All the peepshows now sell substitutes—dolls to have sex with, vibrators, plastic vaginas and penises—and as one woman groused in New York, 'It won't be long before customers can buy a robot from the drug-store and they won't need us at all.'"

]]]]] Paid Sex Surrogates as Therapy

> Sex surrogacy isn't for everyone, but it seems that trying to resolve serious sexual dysfunctions just by talking them over is like learning to drive a car by reading about the history of automobiles. You have to practice.
>
> —Randy Lyman[23]

In 1970, William Masters and Virginia Johnson published their pioneering book on sex therapy, *Human Sexual Inadequacy*. They had developed a successful method of treating sexual dysfunction by suggesting that patients be engaged physically rather than just verbally, thereby creating the basis for modern sex therapy. Their method was to work with the couple, teaching both partners about their bodies and their sexualities. In this book they also described their successful treatment of single men. The only difference between their couples method and the approach required to help single men was that in the latter case the place of the woman in a couple was taken by a surrogate partner, thereby enabling men who did not have partners available to participate in the Masters-Johnson couples-therapy programs. The practice of professional sex surrogacy deserves a place in this chapter, because it is another example of paying for sex, albeit for reasons that are different from those that encourage people to employ the services of prostitutes. Yet the reasons for hiring surrogates to help treat sexual dysfunction will, with time, add to the reasons for indulging in sex with robots.

Surrogate therapy is a three-way process, with many of the sessions involving the client, the surrogate partner, and the patient's therapist. It is the therapist who decides when the client is ready to work directly with the surrogate on their emotional and sexual problems, who introduces the client (usually male) to the surrogate (with the therapist present), and who consults with the surrogate when the therapist feels that the client is ready for intimate and private contact with her. And while the client is attending sessions with the surrogate, he is still being counseled by his therapist, who is also in regular contact with the surrogate.

The treatment process is designed to develop the client's skills at physical and emotional intimacy. All of the most common sexual dysfunctions and their causes can be treated by surrogate-partner therapy, including premature ejaculation, nonconsummation of a relationship, erection difficulties, performance anxiety, and fear of intimacy. The surrogate and the client typically progress through a series of "structured exercises in relaxation, introspection, communication, nurturing, and sensual and sexual touching."[24]

Sex surrogacy is bound to be controversial, because it involves sex as a paid activity. But physical sexual activity is only a relatively small part of the surrogate's typical duties during the therapy process. Raymond Noonan, whose thesis for his master's degree is the standard work on sex surrogacy, surveyed 54 sex surrogates and found that the average surrogate spends approximately 34 percent of the session time talking with the client, in order to provide sexual information, reassurance, and support. Almost half of the time (48.5 percent) is spent on experiential exercises that involve the body, but in a nonsexual way, teaching the client how to feel, how to be aware of the sensory input during sexual encounters. Only 13 percent of the session time is typically spent on physical sexual activities: intercourse, oral sex, and sexual techniques.

In regard to the controversy that attaches to sex surrogacy, Noonan emphasizes that although "the use of surrogates remains controversial, with complex legal, moral, ethical, professional and clinical implications, . . . when performed under the supervision of a licensed therapist, [surrogacy] is completely legal throughout the U.S."[25] And in the online magazine *InnerSelf*, Barbara Roberts points out that "the fact that money is paid for the services of a prostitute, a sexual surrogate or a sex therapist is not the issue. We live in a society where monetary exchange for goods and services is the rule. The intent of those who insist on comparing sex surrogate assisted sex therapy with prostitution is to demean and discredit both. It is a reflection of our basically repressive culture regarding sexuality."[26]

As a profession within the therapy profession, sex surrogacy has never taken off in a big way, though it does boast its own professional association, the International Professional Surrogates Association (IPSA), with its own code of ethics regarding the welfare of both client and surrogate. It appears that in 1977 Masters and Johnson abandoned the recommendation of sex surrogacy, most probably because of a severe nationwide lack of surrogates. Noonan estimated that in 1983–84 there were only about three hundred surrogates practicing in the United States, most of whom were in California and most of the others on the East Coast, but despite the small number, this appears to have been a peak time for the profession, partly because of the subsequent fear of AIDS and partly because most therapists are afraid to recommend the use of sex surrogates to their clients in case of an eventual legal action should the client contract a sexually transmitted disease in the process.

One obvious application of sex surrogacy is in the initiation of young men into sex, a task that in Europe at least has often been the remit of a prostitute. Barbara Roberts, who is a practicing surrogate in California, has found that sex surrogacy has begun in a small way to take on this burden:

> In modern Western societies the messages about sex are extremely contradictory and confusing. We have no traditional rites of passage nor meaningful ceremonies to initiate young people into informed adult sexuality. I hoped that my work might establish standards that could help people of all ages have less confusion about sex and intimate relationships.
>
> Much to my professional satisfaction, there were several enlightened parents who paid for a full course of sexual surrogate assisted therapy so that their sons could be initiated into the wonders of their own sexuality. How lucky to have been those young men's girlfriends or wives! I often wished that parents would take that same enlightened view toward sexual initiation for their daughters, but it was not yet the time for that. I predict, however, that this day will eventually come.[27]

Clearly, sex surrogacy has great potential as a method of treatment, because of the caring, sensitive manner in which a good surrogate can approach the client's sexual problems. The UK Sexual Healing Centre in Bedfordshire* has achieved a high degree of success in treating premature ejaculation, erectile dysfunction, and the inability to consummate a relationship, and a lesser though still significant improvement in resolving the underlying psychogenic causes† of per-

---

*At www.icasa.co.uk.
†Causes that originate in the mind or in mental or emotional processes, rather than being of a physiological nature.

formance anxiety and fear of intimacy. But despite the proven benefits of surrogacy, the paucity of human surrogates currently militates against this form of treatment's becoming mainstream. The solution to this problem should not be difficult for the reader to spot. It is to employ robots as sex surrogates, programming them with the necessary psychosexual knowledge, teaching skills, and humanlike sensitivity.

]]]]] The Moral Justification of Paying for Sex

Many people instantly dismiss the idea of paying for sex, often on the grounds that it is in some way immoral, or because of the commonly held view that only sex with someone with whom one shares genuine affection can be a worthwhile and enjoyable experience. The purpose of this chapter has been to highlight a number of morally valid reasons that paying for sex can be justified, and to demonstrate that for those who do pay for sex, whether frequently or rarely, it can be a positive experience even though they *know* that their sex object has no genuine feelings of affection for them. This indicates that those who consider experimenting by having sex with robots should have no qualms on the basis of the robot's presumed lack of affection for them. Even if their robot exhibited no affection, whether genuine or otherwise, this is no reason to assume that the sexual experience will not be an enriching one for the human. And those who doubt the veracity of this assertion can find comfort in the knowledge that their robot *will* be able to exhibit affection for them at any desired level. It will all be in the software.

7 | Sex Technologies

]]]]] Vibrators Are a Girl's Best Friend

▼ Anyone who has doubts that women will find it appealing or even possible to receive the most incredible, amazing, fantastic orgasms, courtesy of sexual robots, should think again. Think vibrators.

The electromechanical vibrator was invented in the early 1880s as a means of fulfilling a task that hitherto had been accomplished by physicians and before them by midwives. It had been recognized for at least two millennia, and described in medical texts going back that far, that women suffer from a variety of complaints particular to their sex, complaints that collectively went under the name "hysteria," from the Greek for "womb disease." It was also recognized that the most efficacious remedy for hysteria was to bring the patient to orgasm, a task that fell to the medical profession. In *The Technology of Orgasm,* a fascinating and comprehensive account of the history of the vibrator, Rachel Maines quotes a 1653 medical text by Pieter van Foreest that recommends the following:

> When these symptoms indicate, we think it necessary to ask a midwife to assist, so that she can massage the genitalia with one finger inside, using oil of lilies, musk root, erocus, or similar. And in this way the afflicted woman can be aroused to the paroxysm.*

*"Paroxysm" was a term formerly employed for "orgasm."

This kind of stimulation with the finger is recommended by Galen* and Avicenna,[†] among others, most especially for widows, those who live chaste lives, and female religious, as Gradus[‡] proposes; it is less often recommended for very young women, public women, or married women, for whom it is better to engage in intercourse with their spouses.[1]

Why not simply recommend masturbation to women? A very good question. The answer is simply that sexual mores dictated masturbation to be a sin, but it was fine when exactly the same act was performed by a midwife or physician!

Thus, for centuries, the manual massage of women's genitalia was a task frequently undertaken by doctors and midwives, though some physicians of the eighteenth and nineteenth centuries recommended instead horseback riding combined with up to three hours of massage as a method whereby young women could achieve orgasm. All sorts of devices were devised throughout the centuries in attempts to make this task easier and quicker to perform, many of them being manually operated—water-powered and steam-powered devices—which required some measure of skill and effort by their operator. Furthermore, these devices were often heavy, unreliable, and relatively inconvenient to use. Clockwork vibrators, for example, tended to run down rather quickly, and often just at the moment when the woman needed them most, while a steam-driven vibrator invented in the United States in 1869 was inconvenient for doctors to use because they repeatedly had to shovel coal into its boiler.

By the latter quarter of the nineteenth century, physicians had pushed out midwives from this function, realizing that bringing their female patients to "paroxysm" was a nice little earner that added to their regular incomes. It was then that serious demand grew for machines to

---

*A second-century Greek physician, the most famous physician in the Roman Empire.
[†]An eleventh-century Persian physician.
[‡]Gradus, also known as Giovanni Matteo Ferrari da Gradi, was a fifteenth-century Italian physician.

facilitate the task. Many physicians devoted most of their working week to this aspect of their profession, and the number of women a doctor could service using a machine was significantly greater than the number he could cope with manually. Any physician whose consulting rooms boasted a vibration device for this purpose could therefore increase his turnover of patients and hence his income.

The successors to the clockwork and steam generations of vibrators were electrically operated and therefore considerably more effective than their precursors, and once they became available, it was possible through their use for women to experience multiple orgasms. The first electromechanical vibrator was a battery-powered device invented in 1883 by Joseph Mortimer Granville, a British physician. He had previously experimented with clockwork *percuteurs,** "but except for the treatment of neuralgia—and in bad cases of that intractable malady—I do not recommend these instruments."[2]

Granville's annotated drawing of his clockwork device is accompanied by a description of how it is operated. In the illustration, D is the pivot used to wind the clockwork mechanism. When the *percuteur* is wound, a pointed ivory hammer (B) makes percussive movements on the appropriate part of the body, though instead of the ivory point, a flat-headed hammer or brush can be substituted. C marks an ivory button—while this is pressed by the finger, the hammer continues in

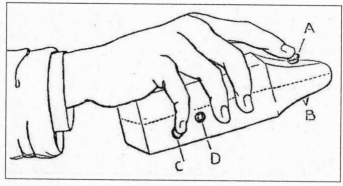

GRANVILLE'S CLOCKWORK *PERCUTEUR*

*Hammers.

action, and when the pressure is released, the hammer stops. The other button, marked A, causes the length of the stroke to be increased and the speed of vibration slightly reduced, while at the same time the force of the hammer blow is augmented.

Granville explains that "the percuteur worked by electricity is, in every way, superior to the clockwork instrument, except as regards portability. In consulting-room practice, the electric instrument answers every purpose most efficiently. The general practitioner will, however, need to provide himself with the clockwork percuteur for use at his patient's house; and, as I have said, although seemingly very weak in its blow, and troublesome, because it requires to be frequently wound, it is by no means ineffectual."

The electromagnetic version of Granville's *percuteur* went into pro-

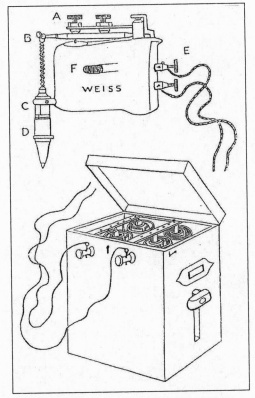

THE *PERCUTEUR* WORKED BY ELECTROMAGNETISM, AND THE BUNSEN'S BATTERY, AS SUPPLIED BY WEISS & SONS

duction in Britain in 1889, manufactured by Weiss & Sons Instrument Manufacturing Company. The terminals of a battery were connected by cables to the vibration device at its terminals, marked E. F was the on/off button. Two screws, marked A, could be adjusted to alter the movement of the hammer, for example by changing its rate of vibration. The screw marked B was for attaching different hammers and brushes to the instrument to create different sensations in the patient. C was a brass cylinder through which the rod of the hammer or brush passed. D was a tube, made of vulcanite,* which was attached with a screw and regulated the length of the stroke made by the hammer.

Accompanying the instrument was a set of hammers and brushes of different shapes, sizes, and purposes, as shown here.

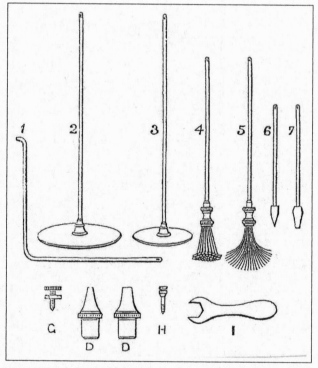

HAMMERS AND BRUSHES EMPLOYED WITH THE ELECTRIC *PERCUTEUR*

*A hard rubber.

There is a bent hammer, marked 1, large and small discs (2 and 3), a hard brush (4) that Granville described as "very effective," a light brush (5) "for relief of superficial pain and to redden the surface," a pointed hammer (6), and a flat-headed hammer (7).

Granville went to considerable lengths to profess that his invention should *not* be employed as a means of sexual relief for women, but instead recommended that it be used only on the muscles of men:

> I should here explain that, with a view to eliminate possible sources of error in the study of these phenomena, I have never yet percussed a female patient, and have not founded any of my conclusions on the treatment of hysterical [fe]males.* This is a matter of much moment in my judgment, and I am, therefore, careful to place the fact on record. I have avoided, and shall continue to avoid, the treatment of women by percussion, simply because I do not want to be hoodwinked, and help to mislead others, by the vagaries of the hysterical state or the characteristic phenomena of mimetic disease.[†3]

Granville further emphasizes his protestations in the conclusion of his book:

> I do not, because I cannot, strongly urge recourse to the method in a considerable number of troublesome afflictions in the treatment of which I have not yet had any large experience of its use. Among these may be mentioned hysteria and the mimetic diseases, and disorders of the sexual organs. . . .

But in the very next paragraph, before going on to recommend the use of his instrument in the treatment of epilepsy, Granville admits

*Granville's book has an unfortunate typographical error here—the word is printed as "males," though the text makes it quite clear that he intended it to be "females."
†"Mimetic disease" (a psychological complaint associated with mimicry) is a term often found linked to "hysteria" in nineteenth- and early-twentieth-century medical writings.

"that the memetic diseases may be successfully treated by nerve-vibration, I have little doubt."

Thus, to all appearances, Granville was distancing himself from the suggestion that his invention could be employed for the sexual arousal and satisfaction of female patients. It does seem inevitable, however, that once Granville had mentioned these possible but nefarious uses of his invention, others would try out these uses. Cynics might therefore suggest that drawing the attention of his medical colleagues to these possibilities in his book was *precisely* what Granville intended with his description of how the machine functioned. Certainly, the medical profession in the United States and other countries was quick to realize the delightful effects that the invention could produce in women, firmly establishing the vibrator as a must-have item for many. Those women who would frequently visit their doctor for sexual relief could now economize by purchasing a vibrator, since the cost was no more than that of a few visits to the doctor.

By the beginning of the twentieth century, vibrator advertisements were appearing regularly in the press. Rachel Maines quotes an explicit advertisement for a five-dollar vibrator from a 1908 issue of the *National Home Journal*:

> To women I address my message of health and beauty. . . . Gentle, soothing, invigorating and refreshing. Invented by a woman who knows a woman's needs. All nature pulsates and vibrates with life.

while a rival manufacturer, the Swedish Vibrator Company of Chicago, advertised its product in the April 1913 edition of *Modern Priscilla* as:

> a machine that gives 30,000 thrilling, invigorating, penetrating, revitalizing vibrations per minute.

In the United States there appears to have been something of a hiatus in the publicity given to vibrators from the 1930s until around

1970, but this might well have been due to a prurient attitude exerting its influence rather than any reduction in their sale and use. By the early 1970s, this attitude had largely worn itself out, and writers on sexual matters had become far less reluctant to extol the virtues of the vibrator. In addition, in 1952 the American Medical Association declared that hysteria is not really an ailment, and since the vibrator would then no longer be used as a medical device, it had to be acknowledged for its real purpose. Furthermore, the advent of 1960s feminism and the accompanying sexual revolution opened up whole new worlds of sexuality for women. Suddenly it was acceptable for women to demand more and better sexual gratification. Thereafter some writers on sex reported on the popularity of achieving sexual satisfaction with the aid of a soft-bristled electric toothbrush (remember Granville's "light brush"!), but in her 1974 book *The New Sex Therapy*, Helen Kaplan expressed no doubts whatsoever and wrote:

> The vibrator provides the strongest, most intense stimulation known. Indeed, it has been said that the electric vibrator represents the only significant advance in sexual technique since the days of Pompeii.

Clearly, one of the strongest sexual trends of the twentieth century was for women to embrace electromechanical devices as an alternative and sometimes more reliable form of achieving sexual satisfaction. And as modern woman has taken an increasingly independent view of her absolute right to enjoy her sexuality to the fullest, so the vibrator has played an increasingly important role in satisfying women's sexual needs. With the advent of the Internet, advertisements can be made absolutely explicit—one company offers a product that gives "clitoral stimulation from 0 to approximately 6,000 oscillations per minute, and vaginal and G-spot stimulation from 0 to approximately 200 rotations per minute." Vibrator sales have soared, partly as a result of the ease and popularity of making purchases from behind the anonymity of the Internet, and in turn this increase in their popularity has made them respectable. In 2005 the largest British pharmacy chain, Boots, belied the traditions of the company's Methodist founders, announcing that it was planning to stock vibrators and place them on its open shelves with no embarrassment.

But while the vibrator was gaining in market acceptance, there were still many who believed and still believe the devices to be the epitome of obscenity. In 1998, for example, the state of Alabama amended its Obscenity Statute, making it "unlawful to produce, distribute or otherwise sell sexual devices that are marketed primarily for the stimulation of human, genital organs." So although the sale of Viagra was perfectly legal in Alabama, achieving sexual satisfaction through the use of certain other products was not. And the penalty for a first offense could be a fine of up to ten thousand dollars and/or one year in prison or one year of hard labor. All this for reaching orgasm in a way that could bring no possible harm to anyone.

Almost immediately after this amendment to the Alabama law came into force, and incensed by its stupidity, four Alabama women, with the support of the American Civil Liberties Union and a few vibrator retailers, filed a lawsuit against the state's attorney general, Bill Pryor, admitting that they had themselves used vibrators "either for therapeutic purposes related to sexual dysfunction, or as an alternative to sexual intercourse." The attorney general argued that vibrators were obscene. The plaintiffs brought forward various expert witnesses, one of whom was Rachel Maines, who testified by affidavit that, inter alia, genital massage technologies "have been available to the citizens of Alabama with or without medical advice and/or supervision, since before the Constitution was written"; that "the FDA* explicitly recognizes massage of the human genitalia as a legitimate therapeutic use of vibrators"; and that the "massage of the genitalia to orgasm has been used as treatment of female sexual problems since the time of Hippocrates, 5th–4th century B.C."[4][†]

---

*The U.S. Food and Drug Administration—the body responsible for controlling all things medical that are sold in the United States.
†The entire affidavit is available at http://www.libidomag.com/nakedbrunch/maines.html.

In deciding on the suit, the court supported Maines's arguments against the 1998 amendment, partly on the basis that "obscenity," the very title of the statute of Alabama law, means something that appeals to "prurient interest, . . . shameful or morbid interest in nudity, sex or excretion." The court found that if the law were to be upheld, then "users of these devices will be denied therapy for, among other things, sexual dysfunction" and that the law "interfered with the very sexual stimulation and eroticism related to marriage and procreation with which the State disclaims any intent to interfere." On this basis, on October 10, 2002, the court overturned the 1998 amendment to the Obscenity Statute, ruling that the law was "overly broad," that it bore "no rational relation to a legitimate state interest," and that it thus violated the due process clause of the Fourteenth Amendment to the United States Constitution.[5]

So vibrator sales are now legal in Alabama *within* the confines of marriage, which is just as well, because their sales are thriving there as everywhere in the United States.* But it is not clear whether the pursuit of orgasm by artificial means would be ruled legal in Alabama for *unmarried* couples or for gay or lesbian couples, nor have I been able to discover any indication that any or all of Connecticut, Georgia, Louisiana, Massachusetts, Mississippi, Nebraska, South Dakota, or Texas have yet repealed laws similar to the Alabama statute that were on their books as of 1998. Bearing in mind the massive sales of vibrators, one can only assume that the law in those states is being broken by huge numbers of women (and even by men, heaven forbid), a conclusion that many will find truly shocking. Enjoying sex? How disgraceful.

In 2003, and undaunted by the Alabama attorney general's convincing defeat in court, the state of Texas attempted to prove once again that the law is at fault, as reported by the *San Francisco Chronicle*:

A Texas housewife is in big trouble with the law for selling a vibrator to a pair of undercover cops, and the Brisbane vibrator com-

*See the next section, "The Popularity of Vibrators: Orgasms on Demand."

pany she works for says Texas is an "antiquated place" with more than its share of "prudes." Joanne Webb, a former fifth-grade teacher and mother of three, was in a county court in Cleburne, Texas, on Monday to answer obscenity charges for selling the vibrator to undercover narcotics officers posing as a dysfunctional married couple in search of a sex aid. Webb, a saleswoman for Passion Parties of Brisbane, faces a year in jail and a $4,000 fine if convicted. "What I did was not obscene," Webb said. "What's obscene is that the government is taking action about what we do in our bedrooms."

The arrest of Webb in Cleburne, a small town 50 miles southwest of Dallas, was the first time that any of the company's 3,000 sales consultants have been busted, said Pat Davis, the president of Passion Parties. She said the company was outraged by the charges and stood behind Webb. "It makes you wonder what they're thinking out there in Texas," Davis said. "They sound like prudes, with antiquated laws. They must have all their street crime under control in Texas if they're going to spend tax money arresting us."[6]

Joanne Webb's troubles were not limited to her being arrested and charged. A few prominent citizens in her hometown of Burleston, citizens with strong Christian beliefs, not only lodged the complaints with the local police that led to her arrest, but they also created trouble for Webb and her husband at local churches, two of which asked the couple to leave. Gloria Gillaspie, a pastor at Lighthouse Church in Burleston, explained that "they didn't want to comply with what was really Christian conduct and that is why they were asked to leave those churches."

Webb was duly charged under a Texas law that allows the sale of sexual toys as long as they are billed as novelties but makes one subject to obscenity charges when they are marketed in a direct manner, showing their sexual role. Webb's lawyer, BeAnn Sisemore, described the Texas obscenity laws as being "so vague that they could be used to prosecute anyone who uses or sells condoms designed to provide stimulation for sexual pleasure." Fortunately, a Texas judge had the good sense to dismiss the case in July 2004, before it could go to trial and waste more of the taxpayers' resources.

]]]]] The Popularity of Vibrators: Orgasms on Demand

In 1976 as few as 1 percent of the American population used vibrators, but in 1982, only six years later, 25 percent of *Cosmopolitan* readers confessed to doing so. More recently, the day after a particular model of vibrator was used by the character Charlotte on *Sex and the City*, stores across North America were sold out of the item. Even those women who have never used one to bring themselves to orgasm cannot deny the popularity of the vibrator, which is being purchased in increasing numbers both on the Internet and in retail stores.*

So the sales of vibrators are booming. The United Kingdom's leading sex-shop chain, Anne Summers, sold 2.5 million in 2004. In Australia, 1 million are sold per year, with 8 million already purchased by early 2005. Americans in 2001 were estimated to be buying 12.5 million

*Twenty-seven percent of all those who responded to the 2004 Durex Global Sex Survey and answered its question on the ownership of vibrators said that they did own a vibrator or an intimate massager. The figure was even higher in both the age groups from twenty-five to thirty-four and forty-five-plus, with more than one-third of respondents being owners. The survey also found, not surprisingly, that vibrators are more popular with women than with men but did not address the question of how many of the male owners used their vibrators on themselves and how many reserved their use for female partners. And as to which countries were shown by the survey to have achieved the highest market penetration for this product category, Iceland led the way with 52 percent of those surveyed, followed by Norway with 50 percent, the United Kingdom with 49, the United States and Sweden both 43, Australia 42, Denmark 41, and China 40 percent. (The lowest usage was found to be in Thailand and Vietnam, with 6 percent and 5 percent respectively.) The statistic for the United States is broadly in line with the results of a survey among more than 1,600 American women, conducted by Knowledge Networks, an independent polling and market-research firm in California. Their survey results were published by the Berman Center and indicated that 51 percent of women in the age group twenty-five to thirty-four had used a vibrator, reducing to between 41 and 46 percent in most other age groups and to 32 percent in the fifty-five to sixties.

vibrators every year, to add to an estimated 50 million plus that were already in the bedrooms of American women at that time,* and by 2005 one of the leading manufacturers, Good Vibrations, estimated that annual sales had risen to 30 million plus.

It is not difficult to understand why vibrators have become so popular with women. The reasons are succinctly, if somewhat drily, explained in a 1996 paper published in the *Journal of Sex Research*, which summarized the opinions of women who used them:

> A majority indicated orgasms triggered by vibrator stimulation were more intense than others. Nearly half experienced multiple orgasms when using a vibrator. Most were very satisfied with their orgasmic experience in autoerotic activity and were either moderately or very satisfied with their orgasmic experience in partnered activity.[7]

]]]]] Vibrations for Men

No matter what use is made by men of female vibrators, the differences between the male and female genitalia obviously call for sex toys for the boys that are different to those made for women.

The first two patented devices designed to help in providing sexual relief for men were both the product of German inventiveness in the early 1950s. The very earliest was the *Gymnastikapparat* (Gymnastic Appliance) designed by Emil Sprenger of Munich, who applied for a patent for his device in May 1949 and had it granted in November 1951.[†]

---

*According to the Web site www.cakenyc.com.
†German patent number 825,137. Sprenger's invention consisted of a hollow cylinder made of glass or some other material, to which an air-evacuation device such as a pump could be connected at one end. The purpose of the invention was described as being to overcome sexual impotence in men, which according to Sprenger's patent application is nearly always based on an inadequate blood supply to the erectile tissue of the penis. To operate the device, the penis was inserted and the air sucked out of the cylinder by the pump, thereby creating a vacuum inside the cylinder. The resulting excess pressure forced blood into the erectile tissue, causing an erection. Springer admitted that on first use the erection may weaken when the container is removed, but he claimed that after repeated use the erection would persist.

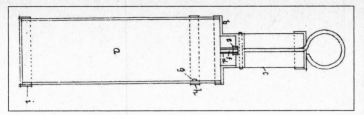

EMIL SPRENGER'S "GYMNASTIC APPLIANCE"

Following closely on Sprenger's heels came Ernst Raeder and Ludwig Hanemann with their *Massageapparat zur Behebung acuter nervös-muskulärer Schwächeerscheinungen* (Massage Device for Relieving Acute Nervous Muscular Debility Symptoms), for which the patent was granted in February 1952.* The symptoms to which the patent title refers are those "arising specially during sexual intercourse."

The first such device developed in the New World appears to be a bagel-shaped penis ring invented in 1966 by Cesareo Barrio of San Leopoldo, Brazil.

The ring was actually a pneumatic or hydraulic chamber with flexible walls. Connected to this chamber by a tube was a pump arrangement that alternately supplied and withdrew fluid from the

---

*German patent number 835,637. The invention was a sleeve made of a watertight and highly elastic material such as rubber, which had a double wall containing sufficient compressed air to the extent necessary for the sleeve to preserve its own shape. To use the device, it is moderately inflated and then slipped over the penis "before commencement of the sex act. The erection which initially occurs is maintained by pumping an appropriate quantity of air into the inner space (marked 14 on the drawing) by means of the rubber bulb (marked 16). . . . On suitable repeated use of the device, this pressure massage at the moment of erection causes a noticeable invigoration of the weakened muscles, so that in due course the massage device will become unnecessary. The desired therapeutic effect is further enhanced by suitable massaging when not performing intercourse."

Clearly, there would be immense practical problems for a man wearing this device on his penis while entering and moving inside his partner and simultaneously operating the rubber bulb in order to maintain his erection. In fact, these difficulties seem so overbearing that one wonders whether this description of the use of the device was not merely a sop to distract prudish German patent officers and whether the intended purpose of the invention was perhaps as a sex machine, human partner not required. This scandalous suggestion might explain the inventors' enthusiasm in recommending "suitable massaging when not performing intercourse."

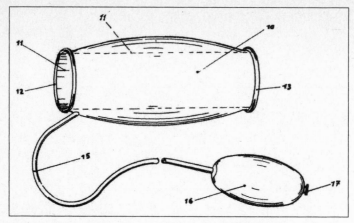

ERNST RAEDER AND LUDWIG HANEMANN'S "MASSAGE DEVICE"

chamber, thereby causing the walls of the chamber to expand and contract, squeezing and relaxing whatever might be in the bagel hole.*

In 1972 a Dutch inventor, Robert Trost, developed a "technological partner" designed to enable the physically handicapped of both

*Barrio's patent document does not make any mention of the word "penis" or any other part of the anatomy. Instead it merely devotes two and a half lines, less than one-fiftieth of the entire text of the patent, to reveal that "one of its principal applications is that of an auxiliary means for the achieving of sexual intercourse in the case of people who are old, paralysed etc."

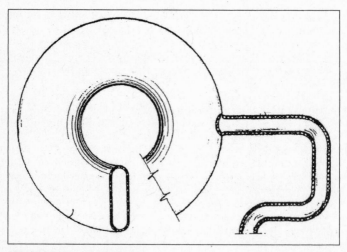

CESAREO BARRIO'S SEX BAGEL

sexes to "attain complete sexual orgasm in an inconspicuous way."[8] The system, called the Coïtron, comprised electrodes that attached to the handicapped person's genitals, which allowed for the adjustment of a pulse generator by means of knobs on a control box. The system was battery-operated, both for portability and for "psychological (fear of electrocution) reasons." By the end of 1972, a working prototype was offered to medical and rehabilitation specialists for further research and testing, and initial results on nonhandicapped men and women were said to be very encouraging. A Dutch Ph.D. student experimented with the Coïtron and wrote his thesis on the basis of these experiments. But the system was never mass-produced because of "the taboos on handicapped people enjoying private sex (i.e., masturbation) which last until today, even in free-thinking Holland."[9]

Another device designed to excite any penis was a gripping system patented by Peter Sobel of Miami Beach in 1975. This invention had attachments covered "with a soft yieldable material, such as rubber or fur" for gently stroking the penis. "The gripping arms of the first and second gripping members are placed on opposite sides of a male genital organ [another patent application that avoids the p word], and one side of the three-way switch is depressed. The variable-speed motor is energized to cause the first and second gripping members to oscillate back and forth and thereby stroke the male genital organ. Again the speed at which the first and second gripping members reciprocate back-and-forth can be gradually adjusted." What fun!*

None of these patented inventions designed for men ever reached commercial viability.* But although the vibrator was invented with

---

*Yet another invention from the 1970s, with a very similar purpose, was a "massaging apparatus," patented in 1976 by Ulrich Glage and his wife, Gisela, of Hamburg, Germany.

The Glages' invention "relates to a device or apparatus for massaging elongated part of the human body, and especially for applying massage to stimulate and enhance the ability for erection." It consisted of a vibrating plastic tube lined with fleshlike rubber that would fit around the entire length of a penis and would operate autonomously or with the added help of the human hand. "The invention provides an apparatus for massage comprising an elongated hollow cylindrical sheath having one closed end and so designed that the outside of the sheath is connected to a vibrating device containing means for the simultaneous generation of two different mechanical vibrations."

LOVE AND SEX WITH ROBOTS

women in mind, and sales of vibrators to women heavily outnumber sales for men, this imbalance has begun to show some signs of a revolution. Since the 1990s, vibration devices have come onto the market designed specifically for men—for example, the Venus line, which was launched in October 1993,[†] from the manufacturer of the Sybian sex machine described later in this chapter.[‡]

A more recent idea, combining penis vibration with synchronized stimulating videos, was launched in December 2004 on the Web site Virtual Sex Machine News,[§] which displays as its banner headline "The Future of Virtual Sex." The site presents an image of what it described as the "Newest Virtual Sex Machine," one that was first announced on the Martin Sargent program *Unscrewed* on the Tech TV network in the United States. This is a suction device with an interface that responds to the activity on screen, allowing the user to watch videos of women porn stars while fantasizing that the women are participating with him in the action. The physical experience generated by the device is thus linked to the visual action by the women, giving the user the virtual-reality experience of having a sexual liaison with a porn star. The operating instructions, as posted on the manufacturer's Web site, represent the height of simplicity:

> Step 1: Put the machine on your penis
> Step 2: Choose any of the girls
> Step 3: Sit back, relax, watch and **FEEL IT!**[**]

]]]]] Artificial Vaginas and Fornicatory Dolls

Artificial aids for assisting in sexual release for men were first mentioned in Japanese literature in the late seventeenth century, during the Genroku era, in a pornographic anthology entitled *Koshoku Tabi-*

---

[*]All these and some eight hundred other sex devices are described in Hoag Levins's 1996 book *American Sex Machines: The Hidden History of Sex at the U.S. Patent Office*, a survey spanning 150 years of sex inventions.
[†]See page 256.
[‡]See pages 253–56.
[§]Available at www.vrinnovations.com.
[**]Their emphasis, not mine.

makura (The Lascivious Traveling Pillow). The device was an artificial vulva, or *azumagata* in Japanese—meaning "woman substitute"—and was made of thin tortoiseshell with an opening lined with velvet to imitate a woman's labia major. In later times *azumagata* were also made of silk and leather and developed into a complete female body called a *do-ningyo*—a "doll body." Paul Tabori provides a description of these fornicatory dolls from the erotic Japanese work *Jiiro haya Shinan* (The Art of Quickly Seducing a Novice).

> A man who is forced to sleep alone can obtain pleasure with a *do-ningyo*. This is the body of a female doll, the image of a girl of thirteen or fourteen with a velvet vulva. But these dolls are only for people of high rank.* Another name of the doll body is even more outspoken: *tahi-joro*—"traveling whore."[10]

Employing fornicatory dolls while traveling became popular in Europe during the late nineteenth century, particularly among sailors. The sexual life of sailors has never been an easy one, living and working as they used to do in an entirely male environment, their trips ashore to the red-light districts of various ports providing just about their only female sexual comfort. Wives and lovers at home could only rarely be visited, so long were the voyages to and from the ships' destinations on other continents. Hence the need for *dames de voyage* (traveling women) as the French called them, and known in Austro-Germany as "sailors' sweethearts." These were dolls in the female form, most often made of cloth and used as sexual outlets by sailors on board ship.

Sex dolls of a less primitive form gained a certain measure of popularity in late-nineteenth-century France.† In 1922 Henry Cary privately published *Erotic Contrivances: Appliances Attached to, or Used in Place of, the Sexual Organs,* a book in which he reproduces and briefly discusses two French advertising circulars, one selling artificial vaginas and the other an entire artificial man or woman.

*Meaning that they were rather expensive.
†See also Bloch's description (in the introduction to part two), pages 178–80.

There is manufactured and sold in Europe today an imitation of the female private parts, even to the pubic hair. These are inflated to give them the desired amount of tightness to the vagina and they are deflated and folded up after using. Circulars describing them usually call them lady travellers, and recommend them for the use of naval officers and others who are deprived of female society for long periods of time. They also advertise that upon receipt of photograph, height, weight and other necessary data, a complete woman will be manufactured to order.

A French circular describes the articles as follows:

"Woman's Belly or Artificial Vagina

"Giving the man the perfect illusion of reality and procuring for him sensations as sweet and voluptuous as those from the woman herself. Outwardly the appliance represents the belly without the thighs. The secret parts, the mount of Venus, covered with abundant and silky hair, the greater lips, the smaller lips, and the clitoris offer themselves to the covetous gaze with rosy colors and temptations as delicious as the pussy of a woman herself.

"In the interior the vagina has wrinkles or folds which embrace and provoke the ejaculation of sperm. The contact is soft and agreeable and the pressure is regulated at will by a pneumatic tube. There is also a lubricating apparatus that is filled beforehand with a warm and oily liquid, and which, under pressure floods the vaginal interior in the same way as the feminine glands secrete at the psychological moment.

"The woman's belly, with lubricator, it is the only apparatus representing exactly the generative organs and capable of giving the effect of reality.

"It can be inflated and deflated at will, and can be folded up and placed in the pocket as easily as a handkerchief.

"The complete apparatus: 100 francs.

"Superior quality: 150 francs."

Other advertisements offer to furnish a complete rubber man, with member of any size desired, and with clockwork mechanism which enables it to perform as desired. Also a woman's torso with generative organs as described in the circular just quoted; also, an entire woman. The latter is made to order, upon receipt of a photograph and measurements, color of hair, and other details, and a perfect likeness is guaranteed, as follows:

"*Complete Body, Artificial Man or Woman*

"All moves, arms, legs, buttocks, head, eyes; a perfect likeness of the person whose photograph is sent. The body in action moves like a living being, pressing, embracing, changing position at will by a simple pressure. The mechanism which gives life to the apparatus is very substantial and cannot get out of order. The complete apparatus, guaranteed against breakage, man or woman, 3000 francs.

"This apparatus can be fitted with a phonographic attachment, recording and speaking at will—man, 3250 francs; woman, 3500 francs.

"In sending photographs of the subject, be sure to give us the height, details of the figure, size of the breasts and buttocks, color of the hair, with sample if possible, and in a word all the information necessary to enable us to complete the figure in an irreproachable manner."

The articles referred to are sold generally throughout Europe, and the fact that the circulars noted come from Paris does not indicate that the French have any monopoly on the traffic. The great bulk of pornographic articles and literature and obscene photographs sold in Europe come from Germany and Austria, the latter country furnishing the most artistic and expensive varieties and Germany, as usual, the cheaper ones.

The popularity of these primitive sex dolls in Europe gave some well-off men the idea of having a doll made in the image of their own lover, past, present, or hoped for. This idea appealed to a few of the surrealist and avant-garde artists of the 1920s, one of whom was Oskar Kokoschka, who had conducted a difficult three-year affair with Alma Mahler, wife of the composer. After their relationship ended, Kokoschka had a life-size doll made in Alma's image by the Munich doll maker Hermine Moos, to whom he had provided a detailed description and some drawings of how he wanted the doll to be made:

> On my drawing I have broadly indicated the flat areas, the incipient hollows and wrinkles that are important to me. Will the skin—I am really extremely impatient to find out what that will be like and how its texture will vary according to the nature of the part of the body it belongs to—make the whole thing richer, tenderer, more human? Take as your ideal . . . Rubens' pictures of his wife, for example the two where she is shown as a young woman with her children. If you are able to carry out this task as I would wish, to deceive me with such magic that when I see it and touch it imagine that I have the woman of my dreams in front of me, then, dear Fräulein Moos, I will be eternally indebted to your skills of invention and your womanly sensitivity, as you may already have deduced from the discussion we had.[11]

Kokoschka bought dresses and lingerie from the best shops in Paris to clothe the doll, and he revealed that when the trunk containing the doll arrived and was being unpacked, his butler became so excited that he had a stroke. But whether Kokoshka actually used the Alma doll for sexual relief appears extremely doubtful, as the doll apparently failed to fulfill his erotic and sexual desires and in the end became no more than a kind of still-life model that in his frustration he destroyed by decapitating it in his garden during a party. He wrote that a Venetian courtesan asked him if he slept with it, but his writings did not answer the question.

Another sad ex-lover who *did* use a lifelike doll as a sex surrogate is amusingly described by Hedy Lamarr in her autobiography. Lamarr was an Austrian-born film actress whose second film, *Ecstasy*, which she made in Czechoslovakia in 1933, shot her to stardom at the age of twenty. This was not because of her acting performance but because she appeared in a nude swimming scene, creating an immediate sensation in Europe and promptly getting the film banned in the United States. Louis B. Mayer was so impressed with her looks that he called her "the most beautiful girl in the world" and took her to Hollywood in 1937, where she embarked on a series of affairs and six marriages that contributed to the considerable unhappiness of her private life.

In *Ecstasy and Me: My Life as a Woman,* Lamarr describes how, when she had discarded Sam, one of her rich lovers, he fell into emotional desolation because their relationship had ended, and had

> a full-sized plastic-rubber doll made to look exactly like me—nude! . . .
>
> The hair looked real, the coloring was accurate (even to the make-up). It had nail polish on the toes as well as the fingers. The figure had obviously been contoured with exquisite care. There was an indecent accuracy to the breasts.[12]

Lamar goes on to explain how she witnessed Sam using his doll, which he named "Hedy-the-Inferior," a use that seemed to provide him with some measure of sexual comfort:

> Sam laid Hedy-the-Inferior on the bed, right in the blue spot.
>
> "Do you love me, darling?" he asked, moving right onto it. He touched those life-like legs, and didn't stop there. I tell you, his master craftsman had included every part of my body.
>
> Sam commenced moving up and down. "Am I hurting you?" he breathed solicitously, "does it feel nice?"
>
> Insane as it was, I couldn't take my eyes off the blue spot!
>
> He was panting, in rhythm. "I love you, I love you, I love." Faster. "I love you," he exclaimed one last time—*"do you love me?"*
>
> I blushed in supreme embarrassment. I knew what was going on the instant he asked that question . . .
>
> And then he was just quivering and whispering to the doll in the blue light.
>
> Finally, he collected himself. He *kissed* those lips, "Thank you darling, you were wonderful. I hope I didn't mess your hair. I know you want to go out tonight . . ."

Despite speculation that the Germans and the Japanese manufactured sex dolls for their armed forces during World War II, no genuine examples appear to have been documented during that period,* but in the mid-1950s a sex toy for men *was* marketed, under the name "Bild Lili." Based on a lewd cartoon character that was popular in Germany at that time, Bild Lili is said to have inspired Ruth Handler in her design for the original Barbie doll.

]]]]] Sex Dolls for the Twenty-first Century

By the early 1980s, blow-up sex dolls were becoming quite big business in some countries but were viewed as obscene in others. In 1982 David Sullivan, a British sex entrepreneur,† attempted to import from West Germany a consignment of inflatable rubber dolls. When inflated, these became life-size replicas of a woman's body, complete with the usual three orifices to provide male customers with sexual gratification. The dolls were seized by the British Customs and Excise as "indecent or obscene articles" and their seizure was upheld in the condemnation proceedings before magistrates and on appeal to the Crown Court. But Sullivan's company, Conegate, then appealed to the High Court in Britain, and, having lost that appeal as well, Conegate appealed yet again, this time to the European Court of Justice, where finally the company won the case in 1987. It turned out that the English law prohibiting the importation of the dolls, which

---

*An article posted on the Internet by Norbert Lenz in 2005 gave an account of "the world's first sex doll," a project initiated by Heinrich Himmler during World War II, with the idea of satisfying the sexual urges of the German troops in France while at the same time keeping the troops away from the disease-ridden prostitutes with whom many of them consorted. This article was taken up by other Web sites and subsequently published by the German newspaper *Bild* and in at least one Scandanavian newspaper. Rather than being of any historic interest, the article was merely an April Fools' Day hoax, and Norbert Lenz is likely a pseudonym. What I find most interesting about this article's publication is that many people believed it, demonstrating that in 2005 there was already a significant measure of belief in the viability of sex robots.

†In the 1970s, Sullivan spotted a gap in the soft-porn market and has since built a $1 billion media empire that includes the newspapers *Daily Sport* and *Sunday Sport*.

dated from 1876, had been superseded by Articles 30 and 36 of the 1957 Treaty of Rome, the document signed when the European Economic Community was created. Under the terms of the treaty, restricting the importation of the dolls into the United Kingdom would have constituted an arbitrary barrier to free trade, and it was free trade that the treaty was specifically designed to promote. The major consequence for the British government of losing this case was that *all* import restrictions on "obscene or indecent" items had to be lifted!

The paucity of published information on the history of sex dolls makes it extremely difficult to date the launch of the first products that appeared on the market in commercially interesting quantities, though the Conegate case indicates that it must have been no later than 1982. Since the mid-1990s at least, various grades of sex doll have been manufactured, ranging from inexpensive inflatable welded-vinyl models, whose looks leave much to be desired but which incorporate an artificial vagina—the main purpose of their customers—through midpriced products made of heavy latex and with convincingly molded hands and feet, imitation eyes in glass or plastic, and styled wigs adorning their mannequin-like heads; up to the top-of-the-line products that in 2006 cost in the region of $7,000, such as the market leader in this price range—RealDoll.

It was in 1996 that Matt McMullen, a California sculptor, revolutionized the sex-toy industry when he launched Nina, the first of a line of products sold under the RealDoll brand name by his company, Abyss Creations. McMullen had previously worked in a Halloween-mask factory, making innocent sculpted female forms in his spare time as a sideline. These were mostly small figures, about twelve inches tall, made of resin and sold as models. With time he began to make larger dolls and to use materials that were softer to the touch. He also designed a skeleton in order to allow his dolls to have limbs that could move.

When McMullen started to advertise his dolls with photographs on his Web site, he received several inquiries from people who believed his products to be sex dolls. When he explained to them that they were wrong, the inquiries changed to ones asking him if he *would* manufacture sex dolls, a group of visitors to his site offering him three thousand dollars each for ten dolls. So he quit his job at the mask factory, developed a silicone material that could be employed to make the doll's genitalia durable and feel right, and by 1996 he was in business.

The RealDoll products are lifelike in appearance as well as being life size and close to life weight. The nine different body sizes advertised on the RealDoll site in early 2006 ranged from five feet one inch tall to five feet ten; they weighed in at between seventy and one hundred pounds; they offered busts from 34A to 44FF, waists from twenty-two to twenty-six inches, and hips from thirty-four to thirty-eight. Other available options included fourteen different female heads, each with its own name: Amanda, Angela, Anna Mae, Brittany, Celine, et al.; seven shades of hair coloring; six different colors for the eyes; fair, medium, tanned, Asian, or African skin tones; and red, blond, or brunette pubic hair that can come shaved, trimmed, or "natural." The dolls are based around articulated skeletons made of steel, have artificial elastic flesh made of silicone, and they come with three functioning "pleasure portals"—vaginal, oral, and anal. Each female doll is thus custom-made, with the buyer able to choose from more than 500 million permutations of these various options.

In addition to the fourteen female models for sale early in 2006, one model of a male doll was also available. It was named Charlie— five feet ten inches tall, with a forty-four-inch chest, a thirty-two-inch waist, and a stocky body. Charlie was priced at $7,000 plus shipping charges and could be provided with "anal entry if desired, plus one size of penis attachment," size not specified. The female RealDolls at that time were slightly less expensive, at $6,500 dollars, and the company was talking of sales in the region of 300 to 350 per year.

RealDoll is by no means the only American brand on the market. A rival California company, CyberOrgasMatrix, uses a different body material—an elastic gel that the manufacturers claim is stronger and more realistic than silicone, as well as being less expensive. Their principal product is the Pandora Peaks model, which, like RealDoll, comes with numerous options. Customers pay according to which options they choose, so that, for example, while vaginal and oral entries are standard, anal entry costs an extra $250. Yet another California manufacturer is SuperBabe, whose doll is modeled on the porn star Vanessa Lace.

The number of sex-doll manufacturers is increasing steadily, as are the Web sites that sell them.* And not to be outdone by the growing band of American producers, companies in China, Germany, and Japan have been getting in on the act. In Nuremberg, Germany, an aircraft mechanic named Michael Harriman claims to have created the world's most sophisticated sex doll, called Andy, with skin made from a silicon-based material employed in plastic surgery, an artificial heart that beats harder during sex, in time with the doll's harder breathing, and internal heaters to raise its body temperature—apart from its feet, which stay cold just as in real life. Andy can be made to move by remote control, wiggling her hips under the sheets and making other suggestive movements, all at the touch of a button. The price is similar to that of the RealDolls, but there are additional charges for special modifications, such as extra-large breasts. Harriman claims that his dolls "are almost impossible to distinguish from the real thing, but I am still developing improvements and I will only be happy when what I have is better than the real thing."

A wide assortment of Chinese offerings is available online and in sex shops, at prices ranging from $50 to $250, as described by Meghan Laslocky:

> *Sweet Spot: A Taste of Things to Come,* a catalogue from Hong Kong, lists nearly 70 different models of blow-up doll, including saucy Sondrine, whose hair, nipples and genitalia glow in the dark; Betty Fat Girl Bouncer, to satisfy the chubby chaser; Brandi Sommer, with "super vibrating LoveClone™ lips"; and The Per-

---

*A comprehensive line of sex dolls and other sex machines is shown, for example, on www.fuckingmachines.com.

fect Date, which is just 36 inches tall and is equipped with a mouth and cup holder built into her head. There's even a dairy maid doll who lactates and has short blonde braids reminiscent of Swiss Miss. Some of the blow-ups vibrate and, oddly enough, scream.[13]

Thus have the sexual lives of sailors, among others, been enriched with the advances in doll design and materials technology, advances that have created realistic skinlike substances such as "cyberskin"* and have thereby made the current generation of sex dolls more comfortable to use than earlier models. That sailors are still avid patrons of such products is of little doubt, and an interesting example of their use was described by Ellen Kleist and Harald Moi in a learned journal in 1993. This report involved the skipper of a fishing trawler from Greenland. After some three months at sea, the skipper had occasion to rouse the ship's engineer in his cabin during the night because of engine trouble. After the engineer left his cabin to sort out the problem, the skipper observed a bump in the engineer's bed, whereupon he found an inflatable doll with an artificial vagina and was tempted into using it in order to assuage his sexual starvation. A few days after this episode, the captain experienced a discharge from his penis, and upon the trawler's return to port in Greenland he sought advice at a hospital in Nanortalik. There had been no women on board the trawler while it was at sea; the skipper denied having had any homosexual contacts, and there was no doubt in the minds of the doctors that the onset of the symptoms was more than two months after leaving port, which meant that the source had to have been on board the trawler. The engineer was then examined by the hospital doctors and found to have gonorrhea. He had observed a mild discharge from his own penis after the ship left port but had not been treated with antibiotics. He admitted having ejaculated into the vagina of the doll just before the skipper had called on him, without washing the doll afterward. He also admitted

---

*Cyberskin is a natural-feeling material that mimics human flesh. It is formed by combining silicone and latex.

having sex with a girl some days before the trawler put to sea. Kleist and Moi's account in *Genitourinary Medicine** suggests that this was the first reported case of the transmission of gonorrhea through an inflatable doll.

The marketing of RealDolls and their cousins from other manufacturers tends to be based on the idea that they are "the perfect woman," perfect because they're always ready and available, because they provide all the benefits of a human female partner without any of the complications involved with human relationships, and because they make no demands of their owners, with no conversation and no foreplay required. And it is precisely because of these attributes, the dolls' lack of "complications" and demands, that they will likely appeal to many of the men who gave such explanations as to why they pay prostitutes for sex, and to others who have similar feelings about their sex lives at home. So, already, in this promotional slant, we can see the basis of the idea that men who use prostitutes should save up their dollars until they can afford a RealDoll. I believe that this will happen in a big way, and that the New York hooker who feared that robot technology would decimate her profession[†] will be proved correct. The signs are already there, as you will soon see.

The most successful manufacturer of sex dolls in Japan is Orient Industries, whose president, Hideo Tsuchiya, started working in the adult sex-aid business in the early 1960s, opening his own store shortly afterward. His business boomed fairly quickly, due largely to two dolls, named Antarctica 1 and Antarctica 2, that achieved media fame of a sort when some of Japan's scientists took them as companions for the winter at Showa Base, Japan's headquarters in Antarctica. At that time Tsuchiya's dolls had permanently open mouths and were inflatable. Although sales were brisk, the blow-up dolls had a tendency to develop leaks and would often burst under the weight of their owners. So Tsuchiya decided that he needed a more durable product, which he achieved by using stronger materials and a design that did not need to

*The branch of medicine dealing with the reproductive and excretory organs.
[†]See page 215.

ONE OF THE ORIENT INDUSTRY DOLLS

be inflated. One of the company's early models, called "Omokage," was specially designed to be dismantled into lower and upper portions, for easy storage in the cramped space of Japan's traditionally small homes.

The growing success of the dolls manufactured by Orient Industries was reported in the *Mainichi Daily News* in late 2003.* "Early on, the showroom was more like a therapy area," recounted Tsuchiya.

*December 11, 2003.

"We'd get old guys who had permission from their wives to buy dolls, or mothers of disabled sons searching for a partner. Nearly all of our customers had some problem related to their sex life."

As the popularity of sex dolls in Japan increased, the attitude of customers toward their dolls transformed, from their being "considered mere instruments in which men could ejaculate to objects of deep affection." By late 2003, Tsuchiya's dolls were selling so well that he was able to boast how it took only ten minutes for fifty new dolls to disappear from the shelves. Standing just under five feet tall and weighing around fifty-seven pounds, the sale of the Orient Industries dolls was beginning to become a fairly big business, increasingly attracting attention from the media. In one report by *Dacapo* [*] journalist Mark Schreiber that appeared in a 2004 edition of *Asian Sex Gazette*,[†] Tsuchiya revealed how "Dutch wives,"[‡] as they are called in Japan, have their own special place and treatment within the confines of Japanese culture, with discarded dolls having the opportunity of funeral rituals redolent of the virtual cemeteries devised for "dead" Tamagotchis.[§]

"A Dutch wife is not merely a doll, or an object. She can be an irreplaceable lover, who provides a sense of emotional healing." Speaking at his showroom near JR Okachimachi Station, where some two dozen of Orient Industries' ersatz females are displayed, Tsuchiya tells *Dacapo*'s reporter that for years his clientele had typically been handicapped men, or single men over forty. But from around six years ago, when he commenced sales via the Internet (www.orient-doll.com), he was mildly surprised to receive a surge of orders from men in their twenties and thirties.

[*]*Dacapo* is a Japanese news digest with a focus on current event feature stories.
[†]April 21, 2004.
[‡]The most authoritative explanation for the origin of the term "Dutch wives" is found in Alan Pate's 2005 book *Ningyō: The Art of the Japanese Doll*. They were originally leather dolls carried aboard Dutch merchant ships, beginning in the seventeenth century; and through their interaction with the Dutch on the trade island of Deshima, established by the Dutch East India company in 1641, the Japanese became familiar with the practice. Pate's own source for this origin was Mitamura Engyo's book *Takeda Hachidai—Eight Generations of the Takeda Family*.
[§]See page 93.

"When I ran my hand along the doll's thigh," confesses *Dacapo*'s reporter, "I felt a shiver of excitement." After observing the painstaking effort that goes into the making of each doll at Orient Industries' factory, the reporter came away enlightened. "Many people might be inclined to disparage sex toys," he writes, "but these dolls truly exemplify Japan's status as a high-tech country!"

Jewel and her sisters are shipped to purchasers in cardboard boxes stamped *kenko kigu* (health apparatus), and users are assured of lifelong after-service. As the vow "until death us do part" may be stretching things a bit, the company anticipates a time when Jewel might outlive her usefulness or her owner. "If a *yome* [bride] is no longer needed, we'll discretely [*sic*] take her off a customer's hands at no charge," Tsuchiya adds. "Twice a year we also arrange for a *kuyo* (Buddhist memorial service) for discarded dolls at the special bodhisattva for dolls at the Shimizu Kannon-do in Ueno Park." Founded in 1631, it's where the "souls" of dolls are consecrated. (Kannon is the Goddess of Mercy.)

A few months after this *Asian Sex Gazette* article appeared, a small group of Japanese entrepreneurs, who had previously been thinking of starting up a regular escort service, decided instead to hire out sex dolls rather than young women. In August 2004 their company, Doll no Mori (Forest of Dolls), opened its first shop in Tokyo's Ota district, specializing in deliveries of dolls to hotels as well as to private homes. Initially Doll no Mori was renting out to around 20 customers per month, but by April 2005 their first shop had increased its customer base to 150 per month, and the business had been franchised to forty other shops nationwide, with monthly turnovers averaging anywhere between $2,500 and $25,000 per shop. The company's manager, Hajime Kimura, explained to the newspaper *Nikkan Gendai* that although "we expected most of the clients to be *eotaku* (geeky) types, as it turned out, most of them are ordinary salarymen in their 30s and 40s." For customers who wish to dress up their dolls, there are optional extras at around $80 each, including wigs, negligees, bathing suits, and other costumes such as school uniforms and French maids' outfits. In a follow-up article in *Nikkan Gendail*,* sex therapist Kim Myung Gun explained, "People have been saying for a long time that men have lost their desire for real women. Rather than have sex with a woman who doesn't fulfill their expectations, they would rather play with something that corresponds to their fantasy, even if she's not real."[14]

It quickly became clear in Japan that the fembot's far less technologically sophisticated ancestor, the sex doll, represents a real threat to the trade of human sex workers. Ryann Connell reported on this growing trend in a 2005 article in the *Mainichi Daily News*: "Rent-a-Doll Blows Hooker Market Wide Open."[15]

Several companies are involved in the bustling trade supplying customers looking to slip it into some silicon[e], with lifelike figurines that set back buyers something in the vicinity of 600,000 yen (about $5,000), as opposed to the simple blow-up types with the permanently open mouths that can be bought from vending machines for a few thousand yen. Prime among the sellers of silicon[e] sex workers is *Doll no Mori*, which runs a 24-hour service supplying love dolls, or "Dutch wives" as the Japanese call them, to customers in southern Tokyo and neighboring Kanagawa Prefecture.

"We opened for business in July this year," said Hajime Kimura, owner of *Doll no Mori*. "Originally, we were going to run a regular call girl service, but one day while we were surfing the Net we found this business offering love doll deliveries. We decided the labor costs would be cheaper and changed our line of business."

Outlays are low, with the doll's initial cost the major investment and wages never a problem for employers. "We've got four dolls working for us at the moment. We get at least one job a day, even on weekdays, so we made back our initial investment in the first month," Kimura says. "Unlike employing people, everything

*April 16, 2005.

we make becomes a profit and we never have to worry about the girls not turning up for work."

Doll no Mori charges start at 13,000 yen (around $110) for a 70-minute session with the dolls, which is about the same price as a regular call girl service. The company boasts of many repeat customers and a membership clientele topping 200. "Nearly all our customers choose our two-hour option."

Within little more than a year after the doll-for-hire idea took root in Japan, sex entrepreneurs in South Korea also started to cash in. Upmarket sex dolls were introduced to the Korean public at the Sexpo exhibition, held in the Seoul Trade Exhibition Center in August 2005. They were seen as a possible antidote to Korea's Special Law on Prostitution that had been placed on the statute books in 2004, and before long, Korean hotels were hiring out "doll experience rooms" for around 25,000 won per hour (around $25), a fee that included a bed, a computer to enable the customer to visit pornographic Web sites, and the use of a doll. This initiative quickly became so successful at plugging the gap created by the antiprostitution law that soon some establishments opened that were dedicated solely to the use of sex dolls, including at least four in the city of Suwon. The owners of these hotels assumed, quite reasonably, that there was no question of their running foul of the law, since their dolls were not human. But the Korean police were not so sure. The news Web site Chosun.com reported, in October 2006, that the police in Gyeonggi province had confirmed that they were "looking into whether these businesses violate the law. . . . Since the sex acts are occurring with a doll and not a human being, it is unclear whether the Special Law on Prostitution applies."

Although the idea of hiring out these dolls appears to be attracting interest from entrepreneurs, the sex-doll industry is still in its infancy and still very much catering to the desires of men, as demonstrated by the fact that of the fifteen models offered on the RealDoll Web site, fourteen are made in the likeness of women and intended for sale to men, while only one is modeled on a man. A likely reason for this disparity—though not the only reason, I'm sure—is that RealDoll's Charlie typically sells for $7,000, and there are far fewer women than men who have thousands of dollars of readily disposable income. But an alternative explanation that has been put forward for the disparity is one with which I strongly disagree—the suggestion that far fewer women than men are interested in using artificial means for getting some or all of their sexual stimulation and for achieving orgasm. Many women claim that the use of sex dolls is very much a "guy thing," but surely such a claim is easily refuted by the widespread use of vibrators among modern women.

]]]]] Sex Machines

The end of the Victorian era witnessed the creation of the first female "self-gratifiers," sex machines that simulated the thrusting movement of a penis inside the vagina. These devices were operated by the woman's turning a handle or pressing on a foot pedal, thereby causing some sort of gear mechanism to move the machine's artificial penis back and forth. The earliest known example of such a device is shown in a sketch dating from around 1900 and incorporated a method of squirting milk into the participant, simulating the ejaculation of semen.

The first commercially available sex machine properly capable of simulating intercourse, and still the most prominent of such machines on the market, was the Sybian,* the brainchild of David Lampert, a former dance instructor in Illinois. In an interview with Jessica West for an article entitled "Plug Into the Ultimate Joy Ride" in *Penthouse Forum* magazine,† Lampert explained what inspired him, in the early 1970s, to devise his robotic penis:

---

*The machine was named after Sybaris, an ancient city of the Greek Empire that was built on what is now the Gulf of Taranto in southern Italy. The city became wealthy, and its inhabitants were reputed to enjoy lives of unrestrained sensual pleasure, providing the origins of the word "sybaritic."
†December 1987.

Over the years, I kept hearing the same complaints from women I met in my dance instruction classes. They were sexually frustrated. Their partners could not, or would not, satisfy them. Some said their husbands had erectile problems due to ill health, age, or indifference. Some of these women confided that they had never experienced an orgasm. That struck me as tragic. I personally could never enjoy sex if the woman is not satisfied.

Lampert's idea became a passion and an obsession. In 1985 he sold his dance studio to devote himself full-time to the development and marketing of what at that time was a revolutionary product.

The Sybian consists of a saddlelike seat containing an electric motor to generate the motion of the machine's phallic "insert." (The inserts come in different sizes and thicknesses and are removable for cleaning.) The Sybian is designed to create two separate movements. The insert rotates within the vagina, and at the same time the area of the Sybian that makes contact with the vulva vibrates, as does the phallic insert itself.

The Sybian is straddled by the woman, who lowers herself onto it

SYBIAN "LOVEMASTER" SEX MACHINE WITH ONE OF ITS INSERTS
ON THE TOP

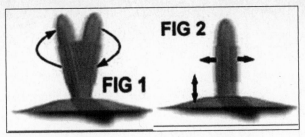

THE ROTATION AND VIBRATION MOVEMENTS OF THE
SYBIAN'S INSERT

when the insert is in place. Separate controls allow for the independent adjustment of the speed of vibration and the speed of rotation. As the Sybian's insert rotates within the vagina, the internal area, including the Grafenberg spot (more commonly known as the G-spot), is stimulated. At the same time, the entire vulva and clitoral area vibrate. The combination of these movements is designed to create a crescendo of orgasms.

After some fifteen years in its development, the Sybian was launched at Couples 87, a weekend convention for sexually uninhibited couples held in St. Louis, Missouri, in the spring of that year. Jessica West describes her reaction on first sight of the machine:

> All eyes were focused on a realistic rubber penis mounted on a vinyl seat. The penis was simultaneously rotating and vibrating at incredible speed. Those were movements no human male could possibly duplicate, at least for any length of time. Just watching that "thing" gyrate made me instantly wet and horny.

Lampert realized that to convince potential buyers of the joys of using his machine, the best method of promotion would be to give women the opportunity to try it out in private, which he did in his hotel suite during the Couples 87 weekend. But before the private sessions began, he gave a demonstration to the trial group as a whole, with their husbands/partners present, in which one of the potential customers had volunteered to be the guinea pig. Jessica West describes what happened after Sally had lowered herself slowly onto the Sybian's insert:

When she was fully seated, Lampert began the demonstration, turning the controls on low speed at first, and then gradually increasing the intensity of both the vibration and the rotation. Sally's face began to contort—first with mild pleasure, then with growing disbelief, and eventually with complete abandon. Her litany of moans and screams, her nails digging into poor Dave Lampert's back, gave ample evidence that she was in the throes of extreme orgasmic excitation. At one point I thought she was going to faint. "Oh, God, oh no, oh yes, don't stop, harder, faster, oh wonderful," she intoned again and again. A communal sigh went up—almost like a communal orgasm. After about 20 minutes of this, it became obvious that Sally could go on coming forever. When Lampert finally turned the machine off, I thought it was probably because his back couldn't take any more. Sally continued to shudder from head to toe for several moments. As she raised herself back on her legs, I could see that her knees were weak. A cheer went up from the crowd, and Sally's husband gently helped her over to a couch. She smiled at her audience like a victorious long distance runner who had reached the finish line, and thanked Lampert for the experience. "All I want to know is, when can I do this again?"

Encouraged by his initial commercial success with the Sybian, Lampert branched out by developing a sex toy for men, an electrically powered "hands-free masturbation aid with controllable stroking action, that gives powerfully satisfying orgasms" and allows the user to "achieve an orgasm in minutes or enjoy sensual stroking for hours." This device was first marketed in October 1993 under the name Venus II, and it was later renamed the Venus 2000. The advertising description of the device suggests that it is easy to use: "Simply insert yourself into a lubricated, flexible, natural gum rubber liner," switch on the machine, adjust the stroke speed to between eight strokes and three hundred strokes per minute, and adjust the stroke length. The company's advertisements point out the obvious benefits of the machine, including that it is "always ready, no partner needed and no risk of disease."

Custom-fit attachments can also be provided. These include a "pump"—a recreational vacuum device that pulls blood into the penis, thereby creating a fully engorged penis from a flaccid state or enhancing an existing erection. Another optional extra is a "head massager," described as an erotic foreplay device that creates a squeeze-release action wherever it is placed.

While the Sybian is perhaps the best known of its kind, there are now many other machines on the market for simulating intercourse. This is hardly surprising in view of Kim Airs's estimate that in 2005 there were seventy-five thousand sex aids on the market, accounting for almost one quarter of the $12 billion taken by the adult-entertainment industry per year.* A different type of machine, with a more thrusting movement of the phallic component, has been the choice of a number of manufacturers, creating products that go under names such as "Stallion," "Invader," "Probe Plus," and "Thrillhammer." The next illustration shows a "Stallion" that, apart from being electrically powered, bears a remarkable similarity in appearance to the *Onanierapparat für Frauen* (masturbation machine for women[†]) in the illustration that follows it, which was built in Germany between the two world wars. This particular example of the German machine had been confiscated by the police, and at one time was exhibited in the Dresden Criminal Museum.[‡] Criminal? Yes, that's right. Certain sexual practices were illegal in Germany during the early part of the twentieth century.[§]

The *Onanierapparat für Frauen* was operated by a foot pedal, which in turn drove a pulley system to push an artificial penis device in

---

*Airs's enthusiasm for the proliferation of sex aids is perfectly understandable—she quit a research position at Harvard in 1993 to found Grand Opening, a woman's sex-toy store, where sales in 2005 were running at around $1 million in each of her two branches.
[†]Also called the "female self-gratifier."
[‡]This image appears on page 604 of the *Sexualwissenschaft* (Sexology) volume of *Bilder-Lexicon* (1930), a German illustrated encyclopedia that described itself as "a reference work for all areas of medical, legal and sociological studies into sex."
[§]Strangely enough, these included the offense of copulation with a statue, which was "classified as a misdemeanor (a public nuisance coupled with indecent exposure) and also claimed compensation in the form of a fine (if the statue was damaged or 'sullied')."[16]

THE "STALLION XL" SEX MACHINE

PEDAL-DRIVEN FEMALE MASTURBATION MACHINE IN THE MAGNUS
HIRSCHFELD MUSEUM, BERLIN

and out of the vagina. The machine was built in 1926 by Russian-Jewish engineers in Leipzig, and passed to Magnus Hirschfeld, a Berlin sexologist and sexual reformer, whose liberalizing activities included his attempts to secure the decriminalization of abortion and homosexuality in Germany. During the period of the Weimar Republic, the era of Christopher Isherwood's *Cabaret*, when almost anything of a sexual nature was socially acceptable in Berlin (even if illegal), Hirschfeld also founded and directed the eponymous Institute of Sexual Science in the Tiergarten area of the city; it operated from 1919 until 1933,* many years before the work of Kinsey and other better-known names in the field of human sexuality research.

Unfortunately for Hirschfeld, he was despised by the Nazis—not only because he was Jewish and gay but also because the Nazis had their own ideas about sex, ideas that made Hirschfeld's sex machine appear to them like some sort of threat. It was "a revolutionary idea, and it was the thing that most upset the Nationalists and the Nazis. The idea of this liberated woman, the Weimar Girl, a woman who could choose her own sexuality."[17] Shortly after the Reichstag fire in Berlin in 1933, and as part of their crusade against Jews, Communists, sex, and anything else to which they took a dislike, the Hitler Youth burned almost all the books from Hirschfeld's institute, as well as his many research files and sex inventions, including the original female self-gratifier.

Many of the sex machines on the market today have been the subject of patent applications. Patent documents make useful reading material for researchers who are interested in how things work, available free of

---

*Hirschfeld opened the institute in July 1919, the first of its kind in the world, attempting to establish sexuality as a science. The institute had a staff of more than forty, working in many different fields: research, sexual counseling, the treatment of venereal diseases, and public sex education. It also hosted the main offices of both the Scientific Humanitarian Committee—the world's first homosexual organization—and the World League for Sexual Reform. From the outset the institute was defamed and denounced by the Nazis as "Jewish," "Social-Democratic," and "offensive to public morals." Hirschfeld eventually fled to France, and the institute was vandalized, looted, and shut down in May 1933.

charge on the Internet, for example, from the U.S. Patent Office.*
Hoag Levins's 1996 book describes many of the U.S. patents for sex
machines granted up to then, but from 1996 up to the end of 2005
almost six thousand additional patents were granted in the United
States alone that contain the word "sexual" in their specification.
Clearly, sex is on the minds of many inventors.

Typical of the sex machines that have been the subject of patent
applications in recent years is one described simply as a "Sexual Aid,"
invented by Larry Taylor of Columbia, South Carolina, and granted on
March 10, 1998.† Rather than attempt to create a more romantic paraphrase of the sexual process than that described in Taylor's patent document, I shall rely here on direct quotations from that document in
order to give readers a taste for the language of patentese, should they
be contemplating further exploration in this field.

As a background to the invention, the patent document wisely
presents near the start an explanation of what happens during sexual
intercourse, just in case the reader has any doubts about the process:

> During sexual intercourse, the penis is reciprocally and slidingly
> received in the vagina. The penis typically does not make substan-
> tial contact with the clitoris. Rather, the abdomen and the transi-
> tion area from which the base of the penis extends provides the
> critical contact that may ultimately lead to sexual satisfaction.
> Sexual satisfaction in females generally is not derived from linear
> translation of the penis through the vagina, rather rhythmic pres-
> sure against and/or frictional engagement with the clitoris.

The background explanation continues by giving the reasons such
inventions are needed:

> Women, for one reason or another, are not always successful in
> finding partners who satisfy their sexual drive. Some women, espe-

---

*At www.uspto.gov.
†U.S. patent number 5,725,473.

cially in view of such lethal sexually transmitted diseases such as acquired immunodeficiency syndrome, or AIDS, prefer to abstain rather than engage in human sexual relations. Although sexual relations may be avoided, whether ill-fated or non-disciplined, sexual drive may not. A need exists for an invention that provides for satisfaction of primal sexual drive yet eliminates reliance on human sexual interaction. Specifically, a need exists for a sexual aid that is adjustable to suit individual needs and provides intimate engagement with and appropriate stimulation of a clitoris.

It is an important, nay, essential part of any patent application to show how and why the invention represents an improvement on earlier inventions. Thus we then find:

Several types of sexual aids are described in the patent literature. Unfortunately, the apparatuses described provide singular excitation means which are received in a vagina in a linear path or engage with a vulva in an arcuately* tangential path.

. . .

Clearly, the above demonstrates a need for a sexual aid providing multiple excitation means that contact the clitoris in a locally arcuate path radially spaced inwardly from the path coincident with a vagina.

Taylor's invention not only overcomes the limitations of earlier inventions by exciting the vagina, clitoris, and anus in "locally arcuate paths," it also provides "multiple excitation means that cyclically contact" the vagina, the clitoris, and the anus and furthermore induces a vacuum phenomenon over a user's nipples.

A succinct description of the machine is given in the patent abstract:

A sexual aid including a housing, mounted on detachable legs and containing a motor that urges a dildo, including vibration means,

*In the form of an arc or bow.

to describe an arcuate path generally coincident with an orifice, such as a vagina. A first stimulator, also containing vibration means, is superposed above the dildo and is urged through an arcuate path concentric with and radially spaced inwardly from that of the dildo, cyclically contacting a clitoris. The sexual aid may include means for introducing a vacuum between the first stimulator and the clitoris. A second stimulator, also containing vibration means, is subjacent the dildo and is urged through an arcuate path concentric with and radially spaced outwardly from that of the dildo, cyclically contacting an anus. The sexual aid includes remotely locatable stimulators that may be placed in contact with a user's nipples and areolae. The sexual aid also provides a vacuum phenomenon between the remotely locatable stimulators and the nipples.

I would be willing to bet you, dear reader, that you had never before thought of sex in these terms. And remember, men, next time you make love and forever in the future, do ensure that you are somehow contacting your lover's clitoris "in a locally arcuate path radially spaced inwardly from the path coincident with a vagina." If you do, she will be yours forever.

]]]]] Virtual Reality and Teledildonics

One way to see VR is as a magical window onto other worlds.

—Howard Rheingold[18]

Virtual reality is a technology that immerses the user in a computer-generated world, perhaps a room, perhaps an underwater city, perhaps a world as enormous as an entire solar system or as small as the inside of part of a human body. Typically the user wears a special helmet and/or goggles and possibly an electronically endowed glove, allowing the exploration of this virtual world in a way that provides realistic feedback to the most crucial senses—sight (in 3-D), sound (in stereo of course), and touch. You are there, right in the midst of your world, controlling and sensing virtual objects in very similar ways to real life, seeing things as they would be if they were real. The technologies employed in virtual-reality systems owe their beginnings to the flight simulators designed for trainee pilots, to the stereophonic sound in our hi-fi systems, and to the 3-D movies such as IMAX that allow the viewer to reach out and almost touch a living dinosaur.

The Virtual Sex Machine described earlier in this chapter is a prime example of virtual reality. The manufacturer's Web site extols the virtues of its product in language that requires no understanding of the technologies involved:

> It strokes your penis with a variable intensity, changes speeds in response to the action on the screen, and grips your penis harder or softer, based entirely on the action shown. It has variable vacuum, and can suck hard or soft, depending on the video that is playing. At various times throughout the scene, again, tied to the movie action, you will feel a stimulation on the tip and body of your penis. . . . In some ways, this device exceeds the ability of a "real" sexual partner, as the sensations are longer and more intense. Not only that, the machine NEVER gets tired.

But virtual reality is not enough by itself to create the whole of this illusion. The sights and sounds of the women on-screen—yes. But the variable-intensity penis stroker—no. That technology falls within a science called "dildonics," meaning computer-controlled sex devices, a word coined by computer visionary Theodor Nelson in his 1974 book *Computer Lib/Dream Machines.* Nelson dreamed up the word "dildonics" in response to the invention of a machine that converted sound into tactile sensation, an "audiotactile stimulation and communications system," patented by a San Francisco inventor, How Wachspress. The title of Wachspress's patent is hardly likely to inspire lustful thoughts, nor would the wording of its abstract, including as it does specifications such as "Control signals are derived from biopotentials or other sources." The patent document describes how the system couples high-pressure sound waves to the skin of its user "for a variety of purposes including sensory substitution, the generation of body music, pleasure stimulation." Sex is simply not mentioned. The document suggests that one use of the device would be to place a probe in the human armpit "for the communication of particular types of messages to the brain without employing the ear," and later in the document we find a description of a probe that "may be inserted in other orifices of the body for a variety of purposes." Clearly, it had not escaped Wachspress's attention that by placing a rounded "coupling device" on the skin of a human body, or in a bodily orifice, sound waves could be converted into vibrations that are sexually stimulating.

Howard Rheingold later explained in his 1991 book *Virtual Reality* the idea of "teledildonics,"* meaning the control of sexual devices via the Internet or like means—simulated sex at a distance:

> Picture yourself a couple decades hence, dressing for a hot night in the virtual village. Before you climb into a suitably padded chamber and put on your 3D glasses, you slip into [a] lightweight bodysuit . . . with the kind of intimate snugness of a condom. Embedded in the inner surface of the suit . . . a mesh of tiny tactile detectors coupled to vibrators of varying degrees of hardness, hundreds of them per square inch, that can receive and transmit a realistic sense of tactile presence. . . . Your partner(s) can move independently in cyberspace, and your representations are able to touch each other, even though your physical bodies might be continents apart. You will whisper in your partner's ear, feel your partner's breath on your neck. You can run your hand over your partner's clavicle, and 6000 miles away, an array of effectors are triggered, in just the right sequence at just the right frequency, to convey the touch exactly the way you wish it to be conveyed. If you don't like the way the encounter is going, or someone requires your presence in physical reality, you can turn it all off by flicking a switch and taking off your virtual birthday suit.[19]

*The word "teledildonics," sometimes referred to as "cyberdildonics," was the creation of Lee Felsenstein during the 1989 Hackers' Conference. It is often wrongly credited to Theodor Nelson.

"Teledildonics," then, is transmitted dildonics. Meredith Balderston and Timothy Mitchell, in their paper "Virtual Vaginas and Pentium Penises: A Critical Study of Teledildonics and Digital S(t)imulation," explain that although this term was originally employed to describe interaction between two people over a distance, it has come to include human-machine sexual interactions. "Today's digital technology is attempting to capitalize on this technological concept, using streaming video, DVDs, real-time chat rooms and remote-controlled sex toys to provide customers with gratifying sexual experiences."[20]

Marlene Maheu explains one of the benefits of teledildonics in her electronic booklet *The Future of Cyber-Sex and Relationship Fidelity*:

> Geographic separation over long periods can often stress a committed relationship and put the relationship at risk for infidelity. Examples of couples who deal with geographic separation include men and women who accept distant work assignments, such as military personnel, scientists, and business people. Virtual contact with electronic devices is likely to be a solution to the loneliness and deprivation caused by long periods of separation. Technology may make the separation more bearable and provide a solution for lonely people away from home. Various devices will allow couples in committed relationships to remain in virtual contact and engage in affection as well as sexual gratification.[21]

One of the technological keys to creating a teledildonic experience is what is called a haptic* interface. Haptic technology allows users to feel as though they are touching something in their virtual world. One example would be the steering wheels used in simulated race-car video games—when the user turns the wheel, the feeling is a simulation of how it would feel to turn a real steering wheel in a real race car at the real speed and on the real racetrack being simulated in the game. Another example is a project at the University of Southern

*Pertaining to the sense of touch.

California, employing two haptic interfaces, one a glovelike device called a CyberGrasp, the other a robot arm called a Phantom. The robot arm is attached to a computer and used as a pointer in three dimensions, just as a mouse is used as a pointer in two dimensions. Motors allow the Phantom to exert a force on a user's hand, giving the feeling of interacting with virtual objects in three dimensions. The CyberGrasp fits over the hand just like a glove and is able to transmit, using a network of artificial tendons, all the sensations felt by a real hand. On one end of an Internet connection, a user of a Phantom robot arm strokes a virtual image of a CyberGrasp glove depicted on his computer screen; on the other end of the Internet connection, the user's partner, wearing a CyberGrasp glove, feels the sensation.

Using a haptic interface to convey hand movements and feelings creates an uncanny effect. Mark Cutosky, a member of Stanford University's Dexterous Manipulation Laboratory, describes the feeling when using a haptic interface to manipulate a robot hand. "Suddenly, it no longer feels like I'm here with my glove and I'm controlling that robot hand over there. Suddenly you feel like that's my hand over there, it's an extension of me."

Haptic Interfaces for Teledildonics

In *Robots Unlimited*, I describe some of the features of the electronic sex surrogate patented by Australian inventor Dominic Choy, a life-size sex doll that is designed to be fully controlled by a computer system. This particular invention is a sexual example of a haptic interface. Choy's invention can be employed in two different versions—in single-user mode the interface connects to a virtual-reality software system that provides all the interactivity; in the two-user mode the haptic interface connects, via the Internet (or similar means), with another haptic interface "worn" by the user's sex partner, allowing the two of them to engage in sex-at-a-distance. When a male user penetrates the artificial vagina in his Choy doll, his partner feels his penis entering her.

Choy's invention represents one form of sexual haptic interface, but one that has the disadvantage of imposing an extra "person," the robot doll, at each end of the transmission. This is fine, indeed ideal, when it is intended to operate the doll in single-user mode, but when two's company, more is a crowd, so other approaches are needed to make sex-at-a-distance less crowded. One such device is the Sinulator, launched in 2004, designed to allow your distant lover to control your sex toy over the Internet. There is a transmitter module that connects to a PC—this measures the speed and force of each thrust of a penis and communicates this data to the software, which translates the data into vibration and pulsing data at the other end. If a man's partner at the other end has her vibrator connected to her Sinulator, the movements of his penis will control the movements of her vibrator.

An alternative method of use allows someone to control a sex toy simply by manipulating the controls of the Sinulator, in much the same way as using a remote-control device for a TV. A semipublic demonstration of teledildonics in action in this way was staged in June 2005 by the New York Museum of Sex. The woman being pleasured was Net Michelle, and the sex machine that was used on her was the Thrillhammer. At the other end of the Internet line, in San Francisco, was Violet Blue, a sex educator, columnist, and author. A Sinulator haptic interface was connected into the chain at both ends, allowing Blue to control the Thrillhammer's thrusts even though the machine was almost three thousand miles away. A camera was set up in the museum for the benefit of the spectators in California. Despite some technical problems before the demonstration got fully under way, eventually Violet Blue did manage to give Net Michelle two transcontinental orgasms, proving that the technology of teledildonics is perfectly viable.

A completely different form of sexual haptic interface is a snugly fitting bodysuit as described by Rheingold. From a psychological perspective, I believe that the bodysuit concept will be more acceptable to the majority of lovers, because even though the suit will require the appropriate artificial genitalia, the experience will bring the lovers closer to each other in the sense that no one else (i.e., no sexual robot) will be between them. And as Maheu explains:

> Body suits will be able to stimulate many different erogenous zones simultaneously, thereby intensifying physical experience.

They will use sensors to stimulate touch and will likely be custom-fitted to accommodate a wide range of body types and proportions. Different sensor pads might be located throughout the body suit, each designed to stimulate a different region of the body in variable and programmable ways.

And so instead of one lover asking the other, "Do you have a condom?" the key question before sex will become, "Is your bodysuit strapped on?" or "Are you connected to the haptic interface?"

]]]]] When?

It is of course extremely difficult to pinpoint the year when robots will be able to accomplish the many tasks described in this book, both tasks of the body and those of the mind. But with the help of a couple of reasonable assumptions, we can make an intelligent estimate. First, the point should be made that the physical aspects of robotics do not have as far to progress on this path as do the mental aspects. There are already robots with very advanced physical dexterity in some fields— for example, Toyota's trumpet-playing robot, which reputedly performs as well as Louis Armstrong; the wrestling robots that take part in the annual All Japan Robot Sumo tournaments; the baseball-batting robot that can process a thousand images each second* of a baseball that has been pitched toward it and accurately hit pitches at speeds of up to a hundred miles per hour; and so forth. The area of research in which most development work remains to be done is artificial intelligence, enabling robots to think, to understand, to be creative, to be able to carry on a conversation, and to exhibit emotion, personality, consciousness, and the many other products of our brainpower. So the question "How long will it be until robots can behave like humans in almost every way?" is very much linked to progress in artificial intelligence.

*This is more than thirty times the number of images processed per second by the human eye.

A broad view of progress in AI during its first fifty years, the period starting with the coining of the term "artificial intelligence" by John McCarthy in 1955, points to progress being somewhat correlated to the great increases we have witnessed in computing speeds and computer memory capacities. It was increases in computing speeds that made possible the progress in computer chess, one of the original aims of AI researchers when they set their science various goals, back in 1956. When chess-playing programs were first developed and tested in competition against one another in the mid-1960s, they were able to analyze something of the order of five chess positions per second, employing methods of evaluating chess positions that considered only a few of the factors that affect human decision making in chess. Deep Blue's success in defeating Garry Kasparov three decades later owed much to its ability to analyze some 200 million moves per second, with an analytical capability that encompassed factors much higher in number and complexity.

Similar or even greater increases in processing speed during the next few decades are inevitable. Electronics experts are continually pushing forward the technologies that determine the speed at which silicon chips can process data, and entirely new computing technologies are being researched for this purpose, technologies that do not rely on silicon and that go under names such as molecular computing, optical computing, and quantum computing. While computer processing speeds continue to increase at dramatic rates, computer memory sizes are also increasing apace. The tiny memory devices you might be using to plug into your laptop, store your music downloads, or enable your camera to retain ever-escalating numbers of high resolution photographs are some of the more obvious signs of the increases in memory capacity that we have witnessed during recent years. The importance of increased memory for tasks in artificial intelligence has been succinctly expressed by Yorick Wilks, one of the world's leading experts for the past three decades on computer translation and other aspects of processing linguistic skills. Wilks says that "Artificial Intelligence is a little software and a lot of data." The reasoning behind this aphorism is that the more knowledge a computer program can access, the more intelligent it will appear to be.

I shall now go out on a limb to some extent and explain why my predicted time line for the development of sophisticated robot companions and lovers focuses on somewhere around the middle of this century as the likely time by when all the robotic goals described in this book will be achieved. My own background in AI is strongly rooted in the field of computer chess, so I'll use that as the basis of my argument. Goethe described chess as "the touchstone of the intellect," a definition that matches most people's perception that to play chess at grandmaster level requires a high degree of intelligence. This is why, when the 1956 workshop on artificial intelligence inaugurated the discipline as an academic division in its own right, those scientists who had come together to define and discuss the goals of AI listed computer chess as one of them. Since then computer chess has often been described by eminent AI gurus as "the drosophila of AI," a reference to the way in which the fruit fly has been seen as the classic test bed for genetics research.

When I first became aware of chess-playing programs that were capable of giving weak human players an interesting game, their standard of play was, in comparison with my own standard, abysmal.* Thirty years later I watched with awe as Garry Kasparov was ripped apart in just one hour, in the sixth and final game of his fateful match against IBM's Deep Blue chess computer. So within those three decades, I had seen progress in this touchstone area of artificial intelligence, from the abysmal to world-shattering. Given that playing chess well is a task that requires much brainpower, I believe that another thirty years from now, give or take a few years, will see strides made in just about every other area of AI, including emotion, personality, and all the mental qualities required of a robot that can behave as you and I do, strides that raise the state of the art in those other branches of AI from where they are now, which in many cases is *better than* abysmal,

*At that time I was Scottish Chess Champion and was soon to be awarded the "International Master" title.

to levels that will exceed those exhibited by the most intelligent, the most capable, the most sensitive, the most loving of humans. All this points to a time around the year 2035. When I asked the noted futurist Ray Kurzweil when he expected the first human-robot marriages to take place, his answer was 2029, but I am somewhat more conservative than Kurzweil in my predictions, and I prefer to add in some contingency to allow for unexpected dips in the enthusiasm for academic research in this area and/or for funding from the traditional sources (i.e., U.S. government agencies) in some of the crucial scientific disciplines. Hence my estimate of midcentury. But there is one major factor that could result in my being proved wrong, a factor that could bring marriageable robots onto the market in Kurzweil's projected time frame or even earlier. That factor is the commercial avarice of big business, particularly in the "adult" sector, and I have witnessed its effects before. When, during the late 1970s and very early 1980s, I used to visit the twice-yearly Consumer Electronics Show in Las Vegas and Chicago, videocassette recorders were just another interesting product item. Within another year or two the market for VCRs, and even more for their bestselling videos (pornography), had expanded beyond all recognition. Sex machines such as the Sybian, the Thrillhammer, and the Stallion are not yet big business, but when their sales reach a certain threshold, watch out! Investment in new product developments might suddenly become available on a massive scale, with an eye to increasing the already astounding profits that the adult-entertainment industry reaps each year. If and when that happens, Kurzweil will be proved right.

]]]]] Erotic Computation Research

I do not believe that it will take more than a decade for sexual applications of artificial intelligence and robotics to become mainstream research topics. When that happens, universities and research institutes will be able to tap into the creative talents of the huge number of students and other researchers who are eager to mix their natural desires with their academic goals. Take a look at this text, from the

Web site of the Erotic Computation Group at MIT, for a taste of university life to come:

VISION

The broad goal of the MIT Media Lab is to explore the impact of modern computing on human society. Groups at the Media Lab study the implications of computational technology at all stages of the human life cycle, from the neurological development of infants and the behavioral learning patterns of children to the sophisticated interaction modalities of adults in digital communities.

Though we are at times reluctant to admit it, all humans are sexual beings. It is time that we overcame the antiquated societal taboos associated with the topic of human sexuality and began to explore it from a critical academic viewpoint. The Erotic Computation Group is devoted to this exciting field of inquiry.

One of the more interesting projects undertaken by the group is that of Dan Maynes-Aminzade, the group's leader, which he calls "Sex Toys of the Future" and which the site describes as follows:

From the simple beauty of the Ibrator to the delicate complexity of the Jackhammer Jesus, sexual appliances have a long history of sophisticated industrial design. Unfortunately, these tools are often ineffective, and their manner of usage is frustratingly crude, having changed little since the first century AD. Dan has a special interest in inventing sexual instruments that understand the artistic intentions of their user, allowing for the enhancement and extension of erotic expression. By designing haptic interfaces that scale gracefully from the sexual neophyte to the experienced professional, he hopes to make sex more accessible, intuitive, and fulfilling.

Any reader wishing to join the group will doubtless be intrigued by its admissions policy:

The Erotic Computation Group seeks creative, hard-working, team-oriented, and sexually competent graduate students. Successful applicants possess varied skills in computer programming, electrical engineering, and mechanical engineering. Special sexual abilities are also important assets. Oral and written communication skills are essential, as our work is regularly presented to visitors and submitted to major conferences and journals.

As I have mentioned earlier, I do not expect it to take more than a decade for such groups to form part of mainstream AI and robotics research. But at the time of writing (early 2007), there are no such groups. The "Erotic Computation Group" is a hoax, perpetrated by MIT students. But what appears on their Web site is all quite believable. In fact, when this first appeared on part of MIT's main Web site, some members of the MIT faculty and staff believed that it was genuine.

8 | The Mental Leap to Sex with Robots

The only unnatural sexual act is that which you cannot perform.

—Alfred Kinsey[1]

▼ In the early years of the twenty-first century, the idea of sex with robots is regarded by many people as outlandish, outrageous, even perverted. But sexual ideas, attitudes, and mores evolve with time, making it interesting to speculate on just how much current thinking needs to change before sex with robots is accepted as one of the normal expressions of human sexuality rather than one of its more bizarre offshoots and for us to ask what the processes will be that bring about such a change. In order to demonstrate the extent to which sexual thinking has altered, particularly during the past century or so, we shall examine changes in attitudes, principally in Britain and America, toward four different aspects of sexuality: homosexuality, oral sex, fornication, and masturbation.

]]]]] Homosexuality

Since Victorian times, no aspect of human sexuality has been the subject of more dramatic changes of attitudes than homosexuality. Ancient Jewish law had prescribed the death penalty for sodomy, based on the biblical teachings of Leviticus 20:13: "If a man lies with a male as with a woman, both of them have committed an abomination; they shall be put to death, their blood is upon them." And since biblical times, several countries and civilizations, including Britain and the United States, have similarly meted out the death penalty for sodomy, as documented by Richard Davenport-Hines, Reay Tannahill, and Gordon Taylor:

Aztec law included the death penalty for homosexuals, male and female; in Peru anyone guilty of sodomy was condemned to be dragged through the streets and hanged, and then burned with all his clothes; the Incas burned sodomites alive in the public square.[2]

In 1627, Pedro Simón, in his *Conquistas de Tierra Firme,* reported on five Italian soldiers, serving in Venezuela, who were "strangled and burnt, with general applause" at the orders of their Spanish commander.[3]

The first Russian state laws against buggery appeared in military statutes drawn up during the eighteenth century reign of Peter the Great. Initially this was punished by burning at the stake, later changed to corporal punishment.[4]

In England the appetite for punishing homosexual behavior with the death penalty appears to have been not completely consistent. An ecclesiastical law of 1290 ordered sodomites to be buried alive, but this sentence seems never to have been imposed, and the few sodomites who were convicted by Church courts were hanged. This was also the punishment prescribed by King Henry VIII in 1533, when a "Buggery Statute" was enacted in Britain, defining sodomy as sexual activity between two men or as bestiality involving an animal and either a man or woman. This brought homosexual behavior, and in particular anal sex, within the jurisdiction of the state courts rather than the ecclesiastical courts as it had been previously, but despite the continuing capital nature of the offense, there were cases in 1541 and 1594 of headmasters who were found to have sexually enjoyed male pupils but survived, not only with their lives but also with their reputations scarcely tarnished.[5]

Following the hanging of the Earl of Castlehaven in 1631, there appear to have been no more executions for sodomy in Britain until the eighteenth century. By the early part of the nineteenth century, executions of homosexuals were steadily increasing—in one English county alone, Middlesex, 28 men were hanged out of a total of 42 convicted sodomites during the period 1805–15. And sodomy was regarded as so base a crime in early-nineteenth-century Britain that in newspaper accounts of the trials and executions of those convicted, it was commonplace to write somewhat euphemistically about their offenses, in contrast to the reports of trials of murder, for which all the gory details would normally be published. In the *Times* of August 13, 1833, for example, the report on the execution of Henry Nicoll,* a retired captain from the Fourteenth Infantry Regiment, says of his crime only that he "was tried and found guilty of an unnatural offense." It was a popular pastime for large crowds to watch executions in those days, and the *Times* reported that "amongst the spectators a large number of females also presented themselves, and, by their shouts manifested their abhorrence of the criminal." The broadside† of Nicoll's execution employs language of an even more venomous kind than was customary for hangings, reflecting the general view of the base level of depravity of his offense, but still without saying what he had done: "Heinous, horribly frightful, and disgusting was the crime for which the above poor Wretched Culprit suffered the severe penalty of the law this morning, Monday, August 12, 1833. . . . Thank heaven the public Gallows of Justice in England is very rarely disgraced by the Execution of such Wretches; but, every person must have observed, with dismay, how greatly the number of diabolical assaults of a similar nature, have lately multiplied in this country."

Equally euphemistic was the *Times*'s wording in reporting on the September 1835 trial at the Old Bailey of John Smith and James

---

*The *Times*, generally an extremely reliable source, gives the spelling as Nicoll, whereas in the broadside it is Nichols.
†Broadsides, later known as broadsheets, were news posters, each one usually devoted to a single news item.

THE TRIAL AND

EXECUTION

OF

CAPT. HENRY NICHOLS,

Who Suffered this Morning,

AT

HORSEMONGER LANE GOAL, SOUTHWARK.

HEINOUS, horribly frightful, and disgusting was the crime for which the above poor Wretched Culprit suffered the severe penalty of the law this morning, Monday, August 12, 1833, at the top of Horsemonger Lane Goal. His name became known to the public during the investigation relative to the death of the unfortunate boy Paviour, who was lately, it is suspected, so inhumanely murdered by a Gang of Miscreants. He was also spoken of, as being concerned with a Captain Beauclerk, who destroyed himself while in Horsemonger Lane Goal, some months ago.

Thank Heavens, the public Gallows of Justice in England is very rarely disgraced by the Execution of such Wretches; but, every person must have observed, with dismay, how greatly the number of diabolical assaults of a similar nature, have lately multiplied in this country.

His TRIAL took place at Croydon, on Friday, August 2, before the Honourable Sir James Parke, when the Jury returned a Verdict of GUILTY, on the clearest possible evidence: the statement of the boy Lawrence, with whom the offence was committed, was very clear. It was confirmed by the testimony of a person connected with Capt. Beauclerk.

The Judge immediately pronounced the dreadful sentence of the Law, advising him to make all the atonement in his power to an offended GOD, the few days he had to live; for, he might rest assured, the sentence would not be mitigated.

Captain Nichols displayed the greatest calmness, and was unmoved throughout the Trial, and even when the dreadful sentence was passed, he remained perfectly collected.

At nine o'clock this morning the full penalty of the law was carried into effect upon the above individual. Upon being brought back from Croydon to Horsemonger-lane he devoted nearly the whole of the time left him in making his peace with his Creator. On the sheriff going to demand his body he was in the chapel with the Rev. Mr. Mann, the chaplain of the goal, and in a firm voice he informed that gentleman that he was satisfied with his sentence, and was fully prepared to die. He trusted that he had found mercy in his God, and that his crime would be forgiven. So firm was his faith, that he did wish to live another hour. In the course of these observations, the rev. gentleman happened to sigh, on which the culprit said, "Do not sigh." Upon seeing Mr. Walters, the governor, he took that gentleman by the hand, and said. "Although an unfavourable impression existed against him in the prison, he had to thank him and the goalers for their tenderness and humanity towards him." He then turned round, and said, "I am ready." On walking along, the bell was being tolled, when he begged that it might be stopped. He then walked with a remarkably firm step to the drop, and in a few minutes he ceased to exist. He apparently died instantaneously. The culprit, who is fifty years of age, was a remarkably fine looking man, and we understand had served in the Peninsular war, and was of a very respectable family. Since his apprehension not a single member of his family ever visited him, and we understand that it is not their intention to demand the corpse. Under these circumstances the body will be sent to one of the hospitals.

Emerson Printer, Today Street.

THE EXECUTION OF CAPTAIN HENRY NICOLL, IN AUGUST 1833.

Pratt, the last men to be executed in Britain for sodomy. Their crime was described simply as "an abominable offense in a house kept by William Bonnell, who was charged as an accessory. The jury returned a verdict of *Guilty* against all the prisoners. Sentence—Death."*

When in 1828 new legislation retained buggery and oral sex as capital offenses in Britain, the number of capital convictions rose to such an extent that during 1842–49 only murder exceeded sodomy as a cause of death sentences, and during one year (1846) there were actually more death sentences passed on those guilty of sodomy than on murderers. However, after 1835 all such sentences were commuted, reflecting the beginnings of a marginally more tolerant attitude in Britain toward homosexuality, and in 1861 the statutory punishment of homosexual behavior was changed from hanging to penal servitude of between ten years and life.

Prior to the American Revolution, British laws were in force in the colonies, and so offenses committed in Virginia, for example, would be subject to those laws and their punishments. Bruce Robinson has chronicled the history of sodomy legislation in the United States, a history paralleling that of Britain. Thus we find that in 1624 a Captain Richard Cornish was charged in Virginia under the British buggery statute with having raped his male servant. He was found guilty, and both he and his servant were hanged. Sodomy became a capital crime in Massachusetts in 1641 (but only between males); the following year Connecticut included sodomy among its twelve capital crimes, and other states soon followed suit. In 1682 the Quaker colony of Pennsylvania became the first jurisdiction in America to make sodomy a noncapital offense. Initially the prescribed punishment was a whipping, a fine equal to one-third of the offender's estate, and six months' hard labor, but in 1700 the punishment was made much harsher—imprisonment for life or castration.

During the years following America's independence from Britain, the death penalty for sodomy was gradually removed from the former

*The record of public executions in Britain for 1835 makes no mention of Bonnell's execution, nor does the *Times,* so presumably he was either reprieved or acquitted on appeal.

colonial laws, though in several states this took quite some time. South Carolina, for example, abolished the death penalty as its punishment only in 1869. And over the next one hundred years, in both the United States and Britain, the prison sentences passed for homosexual behavior gradually grew shorter and shorter, being abolished completely in Britain in 1967.

By the second half of the twentieth century, the punishment of homosexual behavior had also become a rarity in the United States. In 1950 *all* American states had antisodomy laws on their books, and even in 1974, when the American Psychiatric Association crossed homosexuality off its list of diagnoses, the laws of some states still regarded homosexual behaviors as criminal offenses deserving of harsh punishments. Such laws were not dealt a final blow until, on June 26, 2003, after months of public media debate, the U.S. Supreme Court struck down a Texas law banning sexual relationships between gay couples on the grounds that the law was unconstitutional. The basis for that judgment was simply that gay men "are entitled to respect for their private lives." This ruling apparently invalidated laws in four states—Texas, Kansas, Oklahoma, and Missouri—that still prohibited oral and anal sex within same-sex couples, as well as laws in nine other states—Alabama, Florida, Idaho, Louisiana, Mississippi, North Carolina, South Carolina, Utah, and Virginia—that prohibited consensual sodomy (defined for the purposes of those laws to include oral sex) for everyone, homosexual, heterosexual, married or not. Until then the punishment prescribed in Idaho was the harshest remaining in the United States—imprisonment from five years to life.

By the summer of 2003, many of America's leading newspapers, including the *New York Times* and the *Boston Globe,* were publishing announcements of same-sex commitments on their wedding pages. Given the social climate in the United States at the start of the twentieth century, I suspect that anyone at that time suggesting the possibility of same-sex legal commitments within the next hundred years would quite likely have been committed to a lunatic asylum, if not worse.

]]]]] Oral Sex

Many American states had laws that in theory treated oral sex as an act similar to sodomy, an attitude that appeared to have gained currency during the last twenty years of the nineteenth century. In 1900 there were thirteen states with sodomy laws that also encompassed fellatio, and of these there were ten states that had specifically changed their sodomy laws so that they *did* provide for punishing fellatio. By 1920 the number of such states had almost doubled, from thirteen to twenty-four. But despite this legislation, prosecutions were exceedingly rare and the practice was exceedingly popular—by the late 1940s, Kinsey's research indicated that 49 percent of married men in American had performed cunnilingus, while 46 percent of married women had performed fellatio.

Today oral sex is viewed very differently from the way it was eighty years ago. Nowadays it is close to being de rigueur in most heterosexual relationships, with 35 percent of men in the United States and 40 percent of those in France reporting in one study that they had engaged in oral sex (fellatio or cunnilingus or both) during their most recent sexual encounter. For women the percentages were 34 in France and 26 in the United States.

]]]]] Fornication

Another sexual "sin" that we consider in terms of changing attitudes and mores is that of fornication—sexual intercourse other than with one's spouse. The Old Testament prescribes death by stoning for fornication, a punishment that could be meted out, for example, to a woman who was found not to be a virgin at the time of her marriage: "Then they shall bring out the damsel to the door of her father's house, and the men of her city shall stone her with stones that she die: because she hath wrought folly in Israel, to play the whore in her father's house: so shalt thou put evil away from among you." (Deuteronomy 22:21)

The medieval church in England was so obsessed with sex that it absolutely banned all forms of sexual activity other than intercourse between married persons, and then only when carried out with the object of procreating. "In some penitentials fornication was declared a worse sin than murder. In the penitentials of Theodore and Bede the penance imposed for simple fornication was one year, but the penalty was increased according to the frequency of the act and the age and discretion of the parties."[6] And under the influence of the Puritans in England, whose general attitude toward sex is best described as anti, an Adultery Act became law in May 1650 in which the death penalty was stipulated for fornication (and adultery, of course), though it seems that this punishment was never applied.[7] After the Puritans moved from England to the colonies and set up their base in Plymouth, John Demos reports that throughout the seventeenth century they had "a steady succession of trials and convictions for sexual offenses involving single persons. 'Fornication' in particular, was a familiar problem" for which the punishment was "a fine of ten pounds or a public whipping—and applied equally to both parties."

Fast-forward a few hundred years, to 1950s England, when, lo and behold, fornication could still technically be an offense. Gordon Taylor reports on a case in which two unmarried women had spent the night at a hotel with two American soldiers, both couples registering at the hotel as married. The women were prosecuted under the Aliens Order* and were duly committed to prison. But by the twentieth century, such eccentricities of the law had become very much the exception rather than the rule, and sex outside marriage had ceased in most civilized jurisdictions to be a criminal offense.†

---

*A 1920 act of Parliament intended to control, by the use of draconian powers, who could come to Britain and their behavior once in Britain. The term "harmful act" was employed in relation to this act of Parliament, in order to bring before the courts people who had committed a variety of "offenses."

†Amazingly, it was only in January 2005 that the state of Virginia repealed a law dating back to the early nineteenth century, banning sexual relations between two unmarried, consenting, heterosexual adults. In fact, that law had not been enforced since 1847, but it took this long for Virginia to accept, formally, that private relationships should be immune from government interference.

]]]]] Masturbation

> For centuries masturbation or self-stimulation of the genitals has been
> associated with evil and madness, as well as a sin against God, deserving
> death and damnation. Masturbation has been blamed for causing tubercu-
> losis, gonorrhea, epilepsy, and insanity. As recently as 1901 the following
> were included in a list of alleged consequences of masturbation: depravity,
> bedwetting, acne, mental retardation, dull mind, memory loss, hallucination,
> hysteria, psychoses, malaise, weight loss, loss of strength, deafness, blind-
> ness, eye diseases, fever, debility, and sudden death.
>
> —Anne Juhasz[8]

Masturbation is another popular sexual practice that was punished by the church, one most often performed without the collaboration of a (human) partner, just as robot sex will not require the
presence of a human partner. In the early Christian church, priests
would dole out penances* of twenty days' fasting for masturbation, and
even seven days for having a wet dream or "involuntary nocturnal emis-

*A penance is a sacrament in some Christian churches that includes contrition,
confession to a priest, acceptance of punishment, and absolution. In England, until
the nineteenth century, the courts of the established church would punish those
who were found to have committed certain moral offenses such as defamation,
fornication, and adultery. Typical of such punishments is the following order, imposed
by the Lichfield consistory court to the ministers of the parish churches of Walsall
and Rushall and the chapel of Bloxwich, "to call before them Ann Bickley to do
penance for fornication." Ann was required to visit each of the churches on successive
Sundays and "during all the Time of Divine-Service shall stand upon a low Stool
placed before the Reading-Desk, in the Face of the Congregation then assembled,
being cloathed in a white Sheet, in her Stocking Feet, with her Hair about her
ears, and having a white Wand in her hand, and immediately after the End of the
second Lesson the said Ann Bickley shall (with an audible Voice) make her humble
Confession, as follows: WHEREAS I Ann Bickley not having the Fear of God
before mine eyes, but being led by the Instigation of the Devil, and my own carnal
Concupiscence, have committed the Crime of Fornication with William Seney To
the Dishonour of Almighty God, the Breach of his most sacred Laws, The Scandal
and evil Example of others, and the Danger of my own Soul without unfeigned
Repentance; I do humbly acknowledge, and am heartily sorry for this my heinous
Offense, I ask God Pardon and Forgiveness for the same, in Jesus Christ, and pray
him to give me his Grace, not only to enable me to avoid all such like Sin and
Wickedness, but also to live Soberly, Righteously and Godly, all the Days of my Life.
And to that End I desire all you that are here present to join me in saying the Lord's
Prayer, Our Father, etc."

sion." So seriously was masturbation viewed by the clergy that "in the five comparatively short medieval penitential codes, there are twenty-two paragraphs dealing with various degrees of sodomy and bestiality, and not fewer than twenty-five dealing with masturbation on the part of laymen, to say nothing of others dealing separately with masturbation on the part of the clergy."[9]

Jon Knowles's comprehensive history of masturbation, on which much of the following summary of the subject is based, indicates that since ancient times and up to the middle of the twentieth century the practice has almost universally had bad press. For various reasons it was disapproved of by the ancient Greeks and Romans. The early church was opposed to it, as to every other sexual act that could not bring forth children, for which reason Bishop Augustine of Hippo* went so far as to argue that masturbation was an "unnatural" sin and therefore more serious than fornication, rape, incest, and adultery, all of which could lead to pregnancy. Masturbation was a crime in the courts of many European countries during the Middle Ages, and although the offense was rarely discovered and brought to the attention of a court, masturbators could suffer extreme civil penalties, including exile. Emperor Charles V's "Penal Rules" of 1532 even went so far as to establish the death penalty for masturbators (as well as for homosexual behavior and for using contraceptives).

In 1676 the first major work on the evils of masturbation was published: *Letters of Advice from Two Reverend Divines to a Young Gentleman, about a Weighty Case of Conscience, and by Him Recommended to the Serious Perusal of All those that may Fall into the Same Condition.* This publication was the "confession" of a young man who ruined himself through masturbation and saved himself through penance.

During the eighteenth century, the church's perennial opposition to masturbation was taken up by many misguided medical practitioners, and for some 250 years thereafter the medical profession was generally of the view that masturbation was the cause of a plethora of horrible diseases. In 1760, for example, Samuel Tissot promoted the myth that

*A.D. 350–430.

masturbation is evil and leads to "post-masturbation disease."* Physicians in the nineteenth and early twentieth centuries continued to diagnose and treat conditions thought to derive from masturbation, employing "cures" that ranged from food products and diets designed to decrease sexual drive to techniques and devices (such as special chastity belts) that would physically prevent sexual arousal and masturbation. These antimasturbation contraptions included "a genital cage that used springs to hold a boy's penis and scrotum in place and a device that sounded an alarm if a boy had an erection";[10] rings of metal spikes that would stab the penis if it became erect; and metal vulva guards. Around the beginning of the twentieth century, several other techniques came into common practice to keep children's hands away from their genitals, including confinement in straitjackets; wrappings of cold, wet sheets during sleep; applying leeches to the genitals in order to remove blood and congestion that might be created by desire; burning the genital tissue with an electric current or a hot iron; castration; and removal of the clitoris.

At the close of the nineteenth century, the pioneering British sexologist Havelock Ellis became the first authority on sexology to speak out against all this ranting. Fearful of censorship in England, Ellis published his refutation in Philadelphia in 1899, attacking the views put forward by Tissot and his followers. Ellis asserted that they were responsible for the mistaken notions of many medical authorities, notions sustained by nothing more than tradition, and he pointed out that masturbation relieves stress and has a sedative effect.

In contrast to Ellis's liberating teachings, the founder of the Boy Scout movement, Lord Baden-Powell, helped to ensure by his writings in Boy Scout manuals that the fear of masturbation survived well into the twentieth century:†

---

*Tissot's book *L'Onanisme, ou Dissertation Physique sur les Maladies Produites par la Masturbation* (*Masturbation: Physical Dissertation on the Illnesses it Produces*) ran to hundreds of editions, variations, and plagiarized publications, throughout Europe and America, stretching well into the twentieth century and thereby creating a worldwide fear of masturbation that continues to be problematic for young and old alike.
†The following extracts from Boy Scout manuals are reported by Jean Stengers and Anne van Neck.

Smoking and drinking are things that tempt some fellows and not others, but there is one temptation that is pretty sure to come to you at one time or another, and I want just to warn you against it. It is called in our schools "beastliness," and that is about the best name for it. . . . The temptation to self-abuse . . . is a most dangerous thing . . . for should it become a habit, it quickly destroys both health and spirits; he becomes feeble in body and mind, and often ends in a lunatic asylum.

Sometimes the desire is brought on by indigestion, or from eating too rich food, or from constipation. It can therefore be cured by correcting these, and by bathing at once in cold water, or by exercising the upper part of the body by arm exercises, boxing, etc.

It may seem difficult to overcome the temptation the first time, but when you have done so once it will be easier afterwards. If you still have trouble about it, do not keep a secret of it, but go to your scoutmaster and talk it over with him, and all will come right.[11]

Even Sigmund Freud, who acknowledged that masturbation relieves stress and cannot cause a sexually transmitted infection, warned that it could cause certain neurotic disorders and have adverse effects on sexual potency. But despite Freud's warning, sexologists and psychologists increasingly came to agree with Havelock Ellis, and as medical, physiological, psychological, and sexual knowledge advanced in the twentieth century, most authorities dismissed claims that masturbating caused physical ailments, although there were still some who chose to emphasize the possibilities of mental impairment as a result. In fact, the effect of the stigma against masturbation was still strong in the United States in 1937, when studies showed that 90 percent of children caught masturbating were severely threatened, punished, and often terrorized by being told that they would go insane or blind or have their penises cut off or their vaginas sewn closed. At that time 82 percent of college freshmen in America believed that masturbation was dangerous.

Increasingly, however, physicians abandoned the notion that masturbation caused physical or mental dysfunction. In 1950, more than thirty years after its publication in German, Wilhelm Stekel's book *Autoeroticism,* which had suggested that masturbation was universal and normal and that interference with it was the actual cause of problems and disorders, was translated into English. And in 1951, after nearly a half century of warning about the evils of masturbation and the horrors of postmasturbatory disease, the U.S. federal government published *Infant Care,* advising "wise" mothers that it could confuse children who masturbate to tell them that they should desist.

One of the most important results of the work of Alfred Kinsey and his colleagues was the normalization of masturbation and the weakening of the stigma against it. Kinsey's research revealed that of the 20,000 people interviewed during his research, between 92 and 97 percent of the men in his study *Sexual Behavior in the Human Male* and 62 percent of the women in his study *Sexual Behavior in the Human Female* had masturbated. In fact, Kinsey found that it was the behavior in which women most frequently achieved orgasm.

Throughout the United States, the church protested at Kinsey's findings. Even without reading Kinsey's work, Billy Graham wrote, "It is impossible to estimate the damage this book will do to the already deteriorating morals of America," while Senator Joe McCarthy, as was his wont, denounced Kinsey's work as part of the Communist conspiracy. Ultimately, as a result of all this furor, the Rockefeller Foundation withdrew its financial support for Kinsey's research.

Studies after Kinsey's death continued to corroborate his findings. In the late 1960s, his colleague Wardell Pomeroy wrote the books *Girls and Sex* and *Boys and Sex,* advising children about masturbation, reassuring them that "no physical harm can come of it, contrary to the old beliefs, no matter how frequently it is done." In fact, Pomeroy said that masturbation was "a pleasurable and exciting experience. . . . It releases tensions, and is therefore valuable in many ways. . . . It provides a full outlet for fancy, for daydreaming, which is characteristic of adolescence. . . . In itself, it offers a variety that enriches the individual's sex life. . . . It is not only harmless but is positively good and healthy and should be encouraged because it helps young people to grow up sexually in a natural way."[12] Finally, the American medical community pronounced masturbation as normal in the 1972 American Medical Association publication *Human Sexuality*.

Nowadays masturbation is treated as a perfectly normal activity, practiced by the mentally healthy and regarded by the medical profession as being free from any danger of causing mental illness, physical damage, or death. *The Continuum Complete International Encyclopedia of Sexuality* quotes data indicating that about 72 percent of young American husbands masturbate, with an average frequency of about twice a month, and that about 68 percent of young wives do so, slightly less than once a month on average. Possibly as a result of such incontrovertible evidence of the prevalence of masturbation, the Vatican revised its position on the practice in 1992, but only very slightly, with its revision of the *Catechism of the Catholic Church*, suggesting that although masturbation is an "intrinsically and seriously disordered act," the church would in future "take into account emotional immaturity, force of habit, a state of anguish, or other mental or social factors which lessen, indeed even extenuate, the individual's moral guilt."

Thus we can see that over time there have been huge changes in attitude toward all of these sexual practices: homosexuality, oral sex, fornication, and masturbation. Practices that were for centuries treated as very serious, even capital offenses in some of the most "civilized" countries in the world are now widely regarded as thoroughly normal and as leading to fulfilling relationships and satisfactory sex lives. The rates at which such attitude changes have come about have varied, but some of the most dramatic changes have taken only decades rather than centuries.

As with progress in so many other fields, particularly science and technology, progress in social ideas and social change is happening much faster in the first decade of the twenty-first century than it did even fifty years ago, and as with science and technology the *rates* of progress and development in social and sexual ideas are themselves increasing. This will inevitably lead to even more rapid changes in the acceptability of new sexual practices, to the point where blow-up dolls and robots will become widely acceptable within society as our sex partners. And once the sexbot bandwagon starts rolling, nothing will stop it.

]]]]] The Cybersex Era

In two important respects, much of the groundwork has already been laid for the sexual-robot craze to start. First, sexual awareness and experiences are now happening to our children at ever-younger ages, a side effect of the revolution in sexual behavior in the second half of the twentieth century, of the ever-increasing media coverage of sex, and of the availability of pornography and other explicit sexual material on the Internet. The average age of first intercourse in the United Kingdom has fallen from twenty-one for women born in the 1930s to seventeen for those born in 1972. And Ward Elliott, quoting a long-unpublished 1970 Kinsey Institute survey, indicates that 92 percent of married American women who were born before 1900 were virgins at the time of their marriage, a figure that declined, on average, by about 8 percent per decade, to 30 percent for 1950s-born "disco era" women. This change is seen as even more dramatic when measured by the percentages of women who had had premarital sex, for whom the increase was almost ninefold, from 8 percent of women born in the nineteenth century to 70 percent of those born at the peak of the Baby Boom. Mirroring these changes, public tolerance of premarital intercourse has grown markedly since the 1960s. In 1969, 68 percent of the American public thought premarital coitus was wrong; this declined to 48 percent of the general population and only 19 percent of college students by 1975, a gap of only six years.

Just as the youth of today are becoming sexually active earlier than in any previous postwar generation, the age at which children first learn about sex has lowered. Nowadays if a six-year-old tells his classmate that he has just found a condom on the patio, he is just as likely to be asked in reply, "What is a patio?" as "What is a condom?" Given this trend, it is reasonable to assume that society's attitudes on matters sexual will to a significant extent be more and more molded by the attitudes of the younger, sexually active generation.

Another development that lays the foundation for positive changes in attitude to sexual robots is the marriage of sex and technology, a union that started in the closing years of the twentieth century. One hundred years earlier, the invention of the automobile created a splendid venue for lovers lacking privacy, facilitating private assignation and fornication. And much more recently, sex has led some of the most important technological developments within the consumer-electronics industry, being, for instance, the driving force behind the boom in sales of the videocassette recorder (porn videos), then the DVD (more porn), and, of course, the Internet (yet more porn, and the first signs of interactive adult entertainment). These are examples of how social responses to technology sometimes encompass and encourage new sexual behaviors.

These two trends have fused together to create cybersex.* The usage of personal computers has become more and more the province of our youth, a phenomenon that will surely be repeated as handheld PDAs† with wireless connections to the Internet and third-generation mobile phones both become mass-market consumer items for recreational use, including sex-related use. As our youth wholeheartedly embrace such technologies, so sex will increasingly permeate through to their computer screens and the liquid crystal displays (LCD) on their hand-held devices.

]]]]] From Haptic Interface to Sex Robot

When the Web site www.BetterHumans.com conducted a survey in February 2003 to investigate what sex technology most people desire, the clear favorite was "android‡ love slaves" with 41 percent of the votes polled, followed at a discreet distance by mind-to-mind interfaces with 24 percent and virtual-reality sex with 17 percent. Clearly,

*Cybersex is sexual activity or arousal through communication by computer.
†Personal digital assistants.
‡Another term for humanoid.

robots are forming a significant part of the sexual thinking of the technologically aware.

Cybersex is the latest sexual revolution, reflecting both the advances in the technologies that make it possible and the norms and play areas of contemporary sexual culture. Sex has become an activity that instead of simply requiring the physical presence of a second person now appeals to many people in newer and different forms, whether it be the opportunity to meet potential sex partners in an Internet chat room or one of the intimate activities that have been made possible through the development of dildonic and teledildonic devices. In the words of Cheyenne, an online sex-show host,* "Technology has allowed people who may have felt repressed, guilty, unimaginative or just basically sheltered, a way to express their sexuality without boundaries and to explore different sexual worlds."

The sexual possibilities that have been created by the teledildonic age are mind-boggling, so for many people sex today is already rather different, and the differences stem from the technologies. Among the most remarkable of these differences is the lack of a necessity to worry about AIDS and other sexually transmitted diseases, even without a condom, because, through the use of haptic interfaces, sex can now take place between lovers who are in entirely different locations— different homes in different cities, different countries and different continents.

How long it will take for the full potential of these new sexual possibilities to be widely appreciated and adopted is particularly intriguing. As William Ogburn explained in 1964, "Behavioral scientists have long recognized that emerging technology has a powerful influence on human behavior, although frequently there is a delay or lag between the emergence of the technology and the social behavioral adaptation to it." Yet in the case of twenty-first-century sexual behavior, the lag might be minimal.

Before you get carried away with the idea that I intend to suggest that sex between two people will become outmoded, may I state very

*At www.cheyennelive.com.

firmly that I do not believe for one moment that this will happen. What I *am* convinced of is that robot sex will become the only sexual outlet for *a few* sectors of the population—the misfits, the very shy, the sexually inadequate and uneducable—and that for some other sectors of the population robot sex will vary between something to be indulged in occasionally—when one's partner is away from home on a long trip, for example—to an activity that supplements one's regular sex life, perhaps when one's partner is not feeling well or not feeling like sex for some other reason.

The modern era of expanding sexual freedom that began with the sexual revolution of the 1960s takes place in cultural environments typified by dynamic change and increased levels of social tolerance. Commenting in 1978 on some of the effects of this freedom on our view of what is normal in relationships, Maxwell Morris wrote:

> The dawning of a new idealism has given vent to increased sexual vigor and freedom among both sexes. The changing panoramic scenario of sexual liberation may, for example, be illustrated by the increasing number of non-traditional "experiments in living." Innovative living arrangements inclusive of the open marriage, group marriage, unmarried sexual cohabitation, and homosexual cohabitation offer a redefinition of the term "meaningful relationship."[13]

Some of the effects of this same freedom on the sexual aspirations and fulfillment of the individual are described by Dennis Peck in terms of an increase in the potential of our sexual pleasure: "Individual fulfillment through various sexually related activities has resulted in a greater emphasis upon recreational sexual expressions." So in the case of technologically driven sexual practices, the ideas are already with us, even in advance of the general availability of the equipment that will turn these ideas into reality.

In the previous chapter, we discussed some of the haptic interfaces that make computer-driven sex a reality. Now consider this: Assume for a moment that instead of a newly found human lover being at the other end of an Internet link with their own haptic interface, engaging with you in whatever sexual activities your respective hearts desire, there is instead a robot, a sexual robot programmed with the knowledge of countless experienced lovers and all of the world's sex manuals. Would you know the difference?

I believe that this test will be relatively easy for robots to pass, given that the physical feeling you experience will be based on a combination of the physical characteristics of your haptic interface and the skills of your lover. If sex-at-a-distance is physically enjoyable with another person, why does it have to be any less enjoyable if that person is replaced by a robot, as long as the physical connection at your end, your haptic interface, is the same? And if you can enjoy sex-at-a-distance with a robot, then why should you not equally enjoy face-to-face (or however) sex with a robot whose embodiment incorporates all the artificial genitalia and other physical characteristics of your favorite haptic interface, with the added benefits of arms to hold you tight, hands to caress you, and a sexy voice to whisper in your ear? In this transition—from haptic interfaces for human-human sex-at-a-distance to haptic interfaces for human-robot sex-at-a-distance to human-robot sex period—we can see how easy it will be for many people to be converted to the idea of robot sex.

While discussing the physical characteristics of sexually appealing robots, we should not forget the benefits that twenty-first-century design and manufacturing technologies will very soon be bringing to sex dolls and somewhat later to sexbots, benefits that will allow the purchaser or hirer to specify the physical characteristics of the product's genitalia. Women for whom size matters will be able to demand for their malebot any girth and length of penis they desire, while men will be able to choose a fembot's vaginal dimensions to be as tight or as cavernous as they wish. And, of course, on the deluxe models, these dimensions will be changeable with the press of a button, or even by murmuring the right words in the robot's ear. These physical characteristics will represent only some of the popular user features that will be designed into sex robots. Others include all the knowledge in the *Kama Sutra* and similar books, and in the famous Japanese paintings of sexual positions. Just as chess programs are loaded with databases of moves in different chess openings, so the robots can be given databases of different sexual positions and techniques from around the world. It will be possible to set different "levels" or "preferences," in much the same way that different skill levels and style-of-play preferences can be chosen on a chess computer. And the robots will be able to learn what the user likes. On one level a robot could be set to cater, in every encounter, to the user's sexual tastes. Another level could allow for a random choice of sexual activities and/or positions, in order to give the user some surprises. Yet another level could be a "teaching" mode that provides instruction for the sexual novice. By providing a host of different options, manufacturers will make sex robots appealing to just about every sexual orientation and taste.

]]]]] Are Men and Women Different?

Sigmund Freud might have foreseen robot sex as a serious possibility. He used to explain in his lectures that when we dream about complex machines, they always signify the genitals, an explanation that might lead us to speculate that he would have regarded robot sex as little more than the practical implementation of this phenomenon, the replacement of a partner's genitals with the artificial genitals of a machine. If Freud had indeed considered this possibility, how would he have assessed the appeal of sexual robots to men and to women? Would he have predicted that men will be more attracted than women to the idea, or vice versa, or that there would be little or no difference to the sexes in the appeal of sexbots?

There are two major parts to this question. First, do men in general and women in general differ in their sex drives? Second, will men and women be equally likely to embrace the technology of sexual robots, or will men want sex with fembots more than women will want sex with malebots, or vice versa, or any other combination thereof?

The first question is one that has long spawned huge differences of opinion among laypersons and psychologists alike, ranging from William Acton's widely quoted pronouncement in 1857 that "the majority of women (happily for society) are not very much troubled with sexual feeling of any kind," to Barbara Ehrenreich's 1999 article in *Time* magazine, in which she revealed the equally astounding news that woman, not man, is destined to be "the sexual powerhouse of the species." To add to the confusion caused by the highly contrasting views of individual "experts" such as these, four leading textbooks on the subject have innocently combined to create even more doubt as to the truth of the matter: *Our Sexuality*, by Robert Crooks and Karla Baur, is dismissive of the stereotypical view of men as having higher sex drives than women; *Human Sexuality*, the famous tome based on the work of William Masters and Virginia Johnson, acknowledges the existence of the stereotypical view but without coming down for it or against it; *Sexual Interactions* by Albert Allgeier and Elizabeth Allgeier also sits squarely on the fence on this issue; while in *Understanding Human Sexuality*, Janet Hyde and John DeLamater explored the possibility that women might have a higher sex drive than men but failed even to discuss the possibility that the reverse might be true.

It was not until recently that the psychology literature could boast what appears to be a definitive answer to the question, when Roy Baumeister, Kathleen Catanese, and Kathleen Vohs conducted an extremely comprehensive study based on more than 5,400 articles and papers in learned journals and conference proceedings, all of which contributed perceptions on sexual motivation, drive, and desire. Baumeister and his colleagues focused specifically on the desire for sex for its own sake, which is linked very closely to sexual enjoyment— the amount of pleasure we derive from sexual activity.

Baumeister and his group assessed sex drive in terms of the desired frequency of sex, the desired variety of sex acts and partners, the frequency of fantasizing about sex, the frequency of masturbation, the actual number of partners (as opposed to the desired number of partners), the frequency of thinking about sex, the willingness to forgo sex, and the willingness to make sacrifices in other spheres in order to obtain sex. Having surveyed the broad range of available evidence on the relative strength of sex drive in men and women, evidence that was extensive and that came from the diverse methodologies employed by thousands of research psychologists, they came to the following conclusion:

> By all measures, men have a stronger sex drive than women. Men think about sex more often, experience more frequent sexual arousal, have more frequent and varied fantasies, desire sex more often, desire more partners, masturbate more, want sex sooner, are less able or willing to live without sexual gratification, initiate more and refuse less sex, expend more resources and make more sacrifices for sex, desire and enjoy a broader variety of sexual practices, have more favorable and permissive attitudes towards most sexual activities, have fewer complaints about low sex drive in themselves (but more about their partners), and rate their sex drives as stronger than women. There were no measures that showed women having stronger drives than men.[14]

But as Baumeister and his colleagues are quick to point out, this overall conclusion

> does not mean that women do not enjoy sex, nor does it mean that women do not desire sex. It certainly does not mean that women should not desire sex or that they should feel guilty over sexual desire or pleasure. . . . Our conclusion is merely that on average men desire sex more strongly and more frequently than women.

One of the reasons for this disparity is undoubtedly the overzealous sexual demands placed by many men on their women, demands that fail to take into account, for example, the levels of fatigue experienced by many mothers due to their child-care roles, especially if they have jobs as well. Where there is a sexual imbalance in the sense of desired frequency, a robot could be the perfect solution for the enlightened couple.

Let us now address the second of our questions with which we started this discussion: Will men and women be equally likely to embrace the technology of sexual robots? On the basis of Baumeister's conclusion, it would be natural to expect that many more men than women will be enthusiastic about the idea of robot sex and more likely to become customers when sexual robots are on the market at affordable prices. And because technology in general is accepted more slowly by women than it is by men, it might seem likely that women will be slower than men to investigate most aspects of the technology of sex, including sex with robots. On the other hand, the use and history of vibrators and their staggering sales figures suggest that once robot sex gets some good PR from women, this bias will be dramatically reduced and possibly eroded entirely.

Another factor that might increase women's motivation for robot love and robot sex is the recent increase in unwillingness on the part of men to marry. It seems that since men are able nowadays to get sex much more easily than twenty or even ten years ago, they hesitate entering into long-term relationships. This trend will leave a lot of women faced with the prospect of a human lover uncommitted as to the long term. Instead many women might prefer to engage with a sexbot—always willing, always ready to please and to satisfy, and totally committed. This ever-availability of malebots could bring about a dramatic and positive change in the parameters of human love relationships, not necessarily for more sex but rather for sex at the right time.

]]]]] When They Look Like Us

Some people will find it relatively easy to get used to the idea of robots as surrogate humans and alternative sex partners. In general I would expect these to be the technologically aware, those who grow up hand in hand with technology, those whose doubts and questions will relate more to what robots can and cannot do than to their appearance. This sector of society will find pleasure and excitement in exploring the capabilities of robots, including their emotional capacities, their personalities, and their sexual proficiencies and preferences.

Others—possibly because of deep reservations, possibly because of prejudice, possibly because their outlook is so literal that they will need to see realistic humanlike robots before they can come to accept the concept of androids as pseudo peers—will take a lot more convincing. And to convince them, the appearance of the androids will be almost as important as, if not more important than, their technical capabilities.

We have already discussed the general perception among Japanese robot designers, even as far back as the eighteenth-century creators of the *karakuri* tea-serving dolls, that to elicit the most positive reactions from humans, such creations should be humanlike in appearance. This perception can be seen in the way that the Japanese robots of today are increasingly endowed with the physical characteristics of humans. And as the number of domestic robots worldwide grows dramatically, from about 400,000 in 2003 to the UN's prediction of 4.1 million in 2007, so the numbers of robot designers, robot-development companies, and robot-research institutes will mushroom, all fueled by a combination of the money earned from robot sales and government mega-investment. This massive research-and-development effort will rapidly lead to the creation of androids so humanlike in appearance that from a few feet away almost no one will be able to tell the difference. In my view these "waxworks as androids" will be utterly convincing in both their appearance and their movement by 2020, if not sooner.

At the Expo 2005 world exposition in Japan, Hiroshi Ishiguro, a robotics professor at Osaka University, unveiled the most human-looking robot yet. Ishiguro's previous version was called Repliee R1 and had the appearance of a five-year-old Japanese girl. It was made of hard plastic, its head could move in nine directions, and it could make gestures with its arms. The exposition's 22 million visitors, 95 percent of whom were Japanese, were able to see his 2005 creation, Repliee Q1.* She has skin made of a flexible silicone material rather than hard plastic, covering a complex system of forty-two actuators† located in

---

*See page 146.
†Devices that create movement.

the upper part of her body and powered by an air compressor, allowing the gynoid* to turn and react in a humanlike way. Repliee Q1 can flutter her eyelids, she appears to breathe, she can move her hands just like a human, she is responsive to human touch, and she can mimic the human behavior of slightly shifting her position from time to time.

Professor Ishiguro was not under any illusions in 2005 that Repliee Q1 would in its present form pass for a human. But he did believe that that stage in the acceptance of robots will soon be possible. "An android could get away with it for a short time, 5 to 10 seconds. However, if we carefully select the situation we could extend that, to perhaps 10 minutes. More importantly, we have found that people forget she is an android while interacting with her. Consciously, it is easy to see that she is an android, but unconsciously we react to the android as if she were a woman."[15]

Just by looking at Repliee Q1 and comparing her with the stereotypical image of robots as laboratory prototypes, complete with wires and parts hanging out, we can see how quickly progress is being made in this field in Japan. As the android technology of the future combines with developments in haptic sexual interfaces, the first sex robots will start to appear on the market. This might happen as a result of commercial collaboration between the manufacturers of products such as RealDoll and the laboratories in Japan that are leading the way in android research and development. Or it could happen that the Japanese themselves decide they do not need any such collaboration and that, by the way, the "Dutch wives for hire" businesses should be major contributors to robot research budgets. Either way, I do not believe it will be many years before the latest announcements from Japanese robot researchers talk of robots as sex partners and start to demonstrate such capabilities. Why not? The technology necessary for orgasm has been around for a while in the form of the vibrator, and more recently as the Thrillhammer and other similar machines. Think back for a moment to Net Michelle's orgasmic experience, created by

*An android made in the (human) female form. This is not necessarily the same as a fembot, which is usually taken to mean a robot having artificial female genitalia or suitable substitutes in the form of haptic interfaces.

Thrillhammer via a teledildonic interface.* How long can it take for the necessary fusion of technologies to occur?

As the first sexbots reach the market, the publicity for robot sex will take off with a bang. Initial news reports will most likely treat sexbots as a curiosity item, but this will not prevent their existence from becoming widely known. Very quickly, soft-core porn sites and Internet chat groups will start to display and discuss sexbots in action. As more and more people rush to their computer screens to watch others enjoying sex with robots, and as increasing numbers of sexual experimenters are interviewed by a medium anxious to publish all the voyeuristic and vicarious details of the thrills and joys of robot sex, so the mainstream media will stop blushing and cash in on the act. Just as *Marie Claire* published an article in 1994 on the almost unthinkable idea of women paying for sex and enjoying it, so the women's magazines of 2014, if not earlier, will, I am certain, be publishing articles on women's experiences in enjoying sex robots. The semiprivate, semipublic exhibition of teledildonics that took place in 2005 at the New York Museum of Sex can be seen as a pioneering media event in this field. Only a few were able to be present in New York that night, or at the other end of the teledildonic line in San Francisco, but the event was reported in the online version of *Wired,* the highly respected, leading-edge high-tech magazine with a print circulation of more than half a million copies. When such events attract increasing amounts of attention from the mainstream media, albeit as curiosities at first, the idea of sexual robots will quickly spread. The first sexbots to reach the market will be too expensive for most to buy, or even to hire, so for a while these products will be restricted to the upper socioeconomic groups. But this was also true in the very early days of "home cinema" and in the early days of AIBO—Sony's robotic dog. As the media interest in robot sex grows, more people will try the experience, buying and hiring sexbots in numbers sufficient to bring down prices, thereby making sexbots available to men and women from a broader economic spectrum.

*See page 267.

]]]]] Sex Robots for Hire

While much of the initial fascination with sexbots will be prompted by curiosity, it is reasonable to expect some interest to stem from advice given by therapists to their patients. People experiencing psychosexual problems will no longer need and lack the services of human sex surrogates—they can instead be referred to clinics where the surrogates are robots. This does, of course, raise all sorts of ethical and legal questions, especially in the litigious climate of the United States, but setting legal problems aside (and they will exist on a much smaller scale, if at all, in many countries), it seems to me inevitable that sex robots will be employed for therapeutic purposes.

This naturally raises the issue of cyberprostitution. The johns and janes who pay for sex benefit in various ways from their encounters with prostitutes (one form of sex without human love), so they will equally benefit in various ways from sex with robots (another form of sex without human love). And robot prostitutes might become a popular method for people to learn sexual technique before entering into a human relationship. With a robot prostitute, the control of disease is implicit—simply remove the active parts and put them in the disinfecting machine. Cyberprostitutes, on the basis of the fees received for their services, could play an important role in the growth of the robotics industry and the ability of this industry to continually develop more advanced products. Certainly, there are some questions to be answered by the lawmakers of the future regarding robot prostitution. Should it be illegal to have a bevy of robot prostitutes (a robot brothel)? Why should it be, since all current laws apply only to human prostitutes? And if such commercial transactions *are* made illegal, will the Mafia attempt to control the manufacture of sexbots, spreading their availability and making them a source of huge revenues?

]]]]] What Will Our Sex Lives Be Like?

There are obvious social benefits in robot sex—the likely reduction in teenage pregnancy, abortions, sexually transmitted diseases, and pedophilia. And there are also clear personal benefits when sexual bound-aries widen, ushering in new sexual opportunities, some bizarre, oth-ers exciting. In "Impacts of Robotic Sex," Joe Snell pointed to various ways in which robot sex could alter human relationships and human sexuality:

> Techno-virgins will emerge. An entire generation of humans may grow up never having had sex with other humans.
>
> Heterosexual people may use same-sex sexbots to experi-ment with homosexual relations. Or gay people might use other-sex sexbots to experiment with heterosexuality.
>
> Robotic sex may become "better" than human sex. Like many other technologies that have replaced human endeavors, robots may surpass human technique; because they would be programmable, sexbots would meet each individual's needs.[16]

An important aspect of human sexuality is the possibility of fail-ure or denial, making sex and the enjoyment of it somewhat capricious. In order to be better than human sex, the performance of sexbots might need to contain those subtleties of human sexuality that will enable them to mimic this capriciousness. Things that are always great can become boring, but the anticipation, doubt, and hope of each sex-ual experience can be instilled in their human owners if the sexbot is designed with these subtleties.

There are many professions that call for being absent from one's sex partner for varying periods of time. Robots can be the perfect sub-stitutes in these situations, satisfying one's sexual needs without creat-ing any cause for concern about disease and fidelity. For sailors, who a century ago would have been traditional customers for *dames de voy-age*,* a charming female robot would be a great alternative to mastur-bation or a visit to the local brothel when ashore. Ships' pursers will perhaps be loaning them out like library books, instead of administer-ing penicillin jabs for the needy.

*See page 179.

There are many other situations in which a sexbot would be the ideal solution. For those who lose a spouse or a long-term partner, whether to illness, death, or as one of the casualties of a broken relationship, robots could provide the answer. As one ages, it becomes clear that maximal sexual intimacy sometimes takes a very long time to evolve—years, even—and that it redefines itself along the evolution of a loving relationship. Robots will be able to achieve this evolutionary process more quickly than humans, by retaining all the memories of living with their human other, analyzing the relationship characteristics exhibited by their human, and by themselves studying huge databases of relationships and how they are affected by different behaviors, then tuning their own behavior to the needs of their human mate. Humans often do not know what they really want or need, so intuitive robot sex partners are a real requirement, able to discern whether their owner really wants sex or would prefer a nice glass of wine or a walk in the park.

Thought-provoking? Certainly. But far-fetched? Not at all.

> > > > > > > > > >

Conclusion

▼ Imagine a world in which robots are just like us (almost). A world in which the boundary between our perceptions of robots and our perceptions of our fellow humans has become so blurred that most of us treat robots as though they are mental, social, and moral beings. A world in which the general perception of robot creatures is raised to the level of our perception of biological creatures. When this happens, when robot creatures are generally perceived as being similar to biological creatures, the effect on society will be enormous. It will be as though hordes of people from a hitherto-unknown and far-off land have emigrated to our shores, a people who behave like us in many ways but who are very clearly different. These hordes will not pass Shakespeare's sixteenth-century test—"If you prick us, do we not bleed?"—but in most other respects they will appear to be just like us. And their capacity for serving as our companions, our lovers, and our life partners will in many ways be superior to those of mere mortals. I am convinced that this is how the world will be by the year 2050.

The robots of the middle of this century will not be exactly like us, but close. In terms of their outward appearance and behavior, they will be designed to be almost indistinguishable from us to the vast majority of the human population. How will it affect us when we are no longer instinctively able to tell robot from human at a glance? How will it affect the way in which we interact with someone we're meeting for the very first time if we're not certain whether that someone is indeed a someone—or instead a something? Might the differences, such as they will be, between humans and robots create a new form of discrimination? And if so, who (or what) will be the group that is discriminated against? Will we humans be thinking and uttering the phrase "You're only a robot" more often or less often than "You're only a human"? And what will be the impact on society when robots reach a level of sophistication, at which they are able to engender and sustain feelings of romantic love in their humans?

I believe that the social and psychological benefits will be enormous. Almost everyone wants someone to love, but many people have no one. If this natural human desire can be satisfied for everyone who is capable of loving, surely the world will be a much happier place. Many who would otherwise have become social misfits, social outcasts, or even worse will instead be better-balanced human beings. And those who are devastated by the breakdown of their most significant human relationship will be able to speed their emotional recovery by rapidly indulging their desire to love, courtesy of a robot with the appropriately chosen specifications.

Having robots take on the role of partner in relationships with human beings is a natural continuation of the trend in robotics research and development that has already passed through various stages: from industrial robots to service robots to virtual pets to companion and caregiver robots for the elderly. The next stage in this trend is the design and construction of partner robots, sufficiently humanlike and sufficiently appealing in various ways to be considered as our true partners. This might happen first over the Internet, since already there are significant numbers of people falling in love with others they meet in that way, and there is no reason why this phenomenon cannot extend to artificial partners they meet in that same environment, including robots. I expect that many people will find this idea distasteful at first, just as many have found the idea of same-sex love, same-sex physical relationships, and same-sex marriage distasteful. But times change. And just as sexual mores relating to homosexuality, oral sex, fornication, and masturbation have changed so much with time, so attitudes and laws relating to human-robot relationships will similarly develop with time. I believe that the changes in attitudes toward robots will be quite rapid, because social change nowadays happens at a very much faster rate than ever before. This is why I expect marriage with robots to be legalized in some countries by the middle of this century.

Some of the implications of love and marriage with robots raise interesting ethical issues. Although humanoid robots are *artificial* people, will the humans who fall in love with robots somehow reduce the degree of artificiality, by endowing their humanoids with a measure of moral standing? Will it still be so clear that the status of humanoids is firmly "not alive," or will we come to regard them as having a twilight status as "kind of alive," "almost alive," or something similar? As humanoid robots become increasingly sophisticated and increasingly humanlike in their appearance and behavior, the notion "kind of alive" will become an increasingly appropriate epithet for us to apply to these robots, until eventually it becomes almost irresistible to think of them and treat them as being "almost alive." And with the inevitably changing view of what constitutes "almost alive," robots will become regarded more and more as our peers, worthy of our affection, of our love. As a result of this change in perception of the "aliveness" of humanoids, one of the ethical conundrums that will face our children and grandchildren relates to what sorts of rights these robots will deserve. The debate on roboethics has up to now been very much focused on issues that we regard as the unethical *use* of robots. But what about the unethical *treatment* of robots? Should we not in this debate be speaking also on behalf of the robots of the future? I believe that we should. As the idea of humans having robot partners and even marrying robots gains currency, we should consider this prospect not only in terms of what it will mean for society but also what it will mean for the robots when they have consciousness? Today most of us disapprove of cultures where a man can buy a bride or otherwise acquire one without taking into account her wishes. Will our children and their children similarly disapprove of marrying a robot purchased at the local store or over the Internet? Or will the fact that the robot can be set to fall in virtual love with its owner make this practice universally acceptable?

While the topics of love and marriage with robots are certain to lead to much heated debate in the years ahead, still more controversy is likely to be generated by the concept of sex with robots. The rate of development in this field will be even more rapid than the changes in attitudes toward robots, with the advent of sexual robots being a major cause rather than a result of attitude changes. This will be due in part to the enormous sums of money that the developers of such products will be able to reap, and partly because of the enormous worldwide interest in and desire for better sex.

I do not foresee the future, with robots as our partners, as totally without its problems. One aspect that I find scary is not the fact that robots will be able to perform almost any *job* better than the most accomplished human—to be the world's best surgeons, lawyers, politicians, chefs—but that in some ways they will be better husbands, wives, and lovers than our fellow human beings. This raises an important psychological problem, that of our feeling threatened by the possibilities of human-robot sexual interactions because there is a better husband/wife/lover—better in the bedroom, at least, and readily available for purchase for the equivalent of a hundred dollars or so. This fear is a natural reaction, akin to the common response that many people still have to sex toys, and to vibrators in particular. Many straight men, for example, feel that a vibrator is a threat to them, believing it could replace them as their woman's preferred sexual partner. Such a feeling is not helped by our contemporary sexual culture, in which the need for a man to be able to sexually please and satisfy his woman is promoted so widely in books and other media and is often the subject of boastful conversation. Put simply, most men would feel inadequate if they believed that their woman enjoyed better orgasms courtesy of a vibrator or a robot than those that the men themselves could provide on a regular basis. But in the final analysis, I believe that one of the great benefits of sexual robots will be their ability to teach lovemaking skills, so that men who do feel inadequate will be able to take unlimited lessons, in private, from robot lovers who possess an unrivaled level of knowledge of sexual techniques and psychosexual problems, combined with great skills as sensitive, patient teachers. This importance of robots, as a means of teaching and enhancing sexual technique should not be underestimated. So many relationships founder because of dissatisfaction in the bedroom, and so many men suffer, as do their partners, because they are unable for whatever reason (including embarrassment) to work to improve their lovemaking skills.

Just as many straight men fear the effect that vibrators might have on their sex lives, so many straight women will deny any need for a vibrator because they already feel completely sexually satisfied by their regular sex partner(s), and for those women it might be the case that whatever additional sexual pleasures robots can offer them will not be of sufficient interest to encourage participation in robot sex on a regular basis. But the psychology literature, both popular and academic, is sufficiently replete with data on sexually frustrated women and the sales of vibrators are burgeoning to such an extent that we cannot doubt the enormous popularity of robot lovers when they become commercially available. And some women will, of course, join the ranks of those who wish to avail themselves of their robot's sexual-teaching skills.

Given that sex with robots will inevitably become a popular human activity, I feel it is important to consider the ethical implications of such encounters—the implications for the human participant in a sexual encounter with a robot, the implications for those (humans) who are emotionally close to the human participant, the implications for society as a whole, and even the implications for the robot participant.

For the human participant, some of the benefits of robot sex create their own ethical justification. The capability of robots to teach all known aspects of sexual technique will turn receptive students into virtuoso lovers. No longer will a partner in a human-human relationship need to suffer from lousy sex, mediocre sex, or anything less than great sex. Marriages and partnerships that today are in trouble in the bedroom will no longer be at risk, thanks to the practical instruction in sex that will be available to all. And, having taken as much instruction as is necessary to enable a participant to satisfy their human partner, the student's sexual confidence will be boosted, thereby creating a more balanced human being.

Another advantage of robot sex will come from robots' therapeutic capabilities, in helping those who suffer from psychosexual hang-ups. Most people cannot afford to seek professional help in resolving such problems, and among those who *could* have access to human therapists, a significant proportion are too embarrassed to discuss their sex lives with others. But robots, in addition to being excellent sex teachers, will also be sympathetic counselors, curing far more cases of psychosexual inadequacy than human therapists ever could. Yet another service for the human participant is the possibility of being able to experiment with different sexualities. People who are uncertain of their own sexuality will be able to try out same-sex and opposite-sex robots in total anonymity.

For those humans whose partners are sexually involved with robots, there can be ethical advantages over and above the joy of sex with improving and skilled lovers. The availability of regular sex with a robot will dramatically reduce the incidence of infidelity as we know it today, though some human spouses and lovers might consider robot sex to be just as unfaithful as sex with another person. Or will sexual ethics come to regard encounters with robots as being as innocent as the use of a vibrator is regarded today?

One of the effects on society of robot sex might be that the fundamental difference between the sexual motivations of men and women undergoes a rethink from both the male and the female points of view. As robots become not only our lovers but also our tutors in lovemaking skills, they could enhance the sexual experience for many men and women by increasing, for each sex, the benefits and feelings most often appreciated by the opposite sex. Sexual robots could encourage men toward a deeper exploration of their feelings of emotional attraction to their human sex partners, while women could be encouraged and taught how to achieve greater physical pleasure from sex. The result of all this tuition would be higher levels of pleasure and satisfaction, both for men and for women, not only in their couplings with robots but also in their human sexual encounters.

For society as a whole, there are clear social and ethical benefits in making sexbots available to those who cannot refrain from indulging in illegal and antisocial sexual practices. In some cases the sexbot will be able to provide the satisfaction necessary to assuage the craving and whatever therapy is necessary to cure the underlying problem that causes the illegal or antisocial practice. And in any event the human will no longer be breaking the bounds of social convention or breaking the law.

Because of all these positives, I believe that for the vast majority of the human population sex with robots will come to be regarded as ethically "correct," as a good thing. But the ethics of robot sex is a very broad subject, creating many different problems for the lawmakers who will come to draft the legislation that converts popular ethics into laws. First, there is the question of how one's use of one's own sex robot will affect other people—one's spouse or partner in particular— and not only because of the question of whether sex with a robot will be considered unfaithful. Will it be unethical in some way to say to one's regular human sex partner, "Not tonight, darling. I'm going to make it with the robot"? (Some couples will, of course, own two robots, a malebot and a fembot, and will enjoy orgiastic sessions in which three or all four of them take part.) And how about robot swapping? Will it be viewed as similar to wife swapping?

There are also issues relating to the use of other people's sexbots. What will be the ethics of lending your sexbot to friends, or borrowing theirs? What about using a friend's sexbot without telling the friend? And there will certainly be ethical (and legal) issues relating to the use of sexbots by minors. Should the age of consent for sex with a robot be the same as that for sex with a human? And what about the ethics of an adult's encouraging a minor to have sex with a robot? Will it be regarded as a sex-educational experience or as a corrupting influence? And how will ethicists and lawyers deal with parents when one parent wants their child to have sex lessons from a robot but the other does not?

Finally, there is the matter of the ethics of robot sex as it affects the robot itself. When robots are so highly developed that without an inspection of their innards they appear almost indistinguishable from humans, should we assume that simply because they are not biological creatures it is totally acceptable for us to have sex with these objects of our creation whenever we wish? If robots become, for all emotional and practical purposes, surrogate humans, will we not have ethical obligations toward *them*? What happens when a robot's owner feels randy but the robot's programming causes it to shy away, possibly because it is running its self-test software or downloading some new knowledge and does not wish to be interrupted, or possibly because its personality was designed in such a way that it sometimes says no for whatever reason? Under such circumstances is it akin to rape if the robot's owner countermands the robot's indicated wish to refrain from sex on a particular occasion? All these examples warn of a minefield for ethicists and lawyers, which partly explains why "roboethics" is becoming a respectable academic topic, with conferences beginning to spring up, particularly in Europe. So the subject is very much under discussion, although the debate is still in its early stages.

Malebots and fembots will inevitably become huge commercial successes. Initially, much of the enthusiasm for coupling with sexbots will be prompted by curiosity, as our natural urges compel most of us to ignore the ethical debate and strive to discover how robot sex compares with the real thing. Many of those who are robot-sex virgins will get their first experience as a result of wanting to know how it feels, but once this curiosity has been satisfied, and once the initial media exposure has made everyone aware of the pros and cons of the robot sexual experience, the demands of the market will drive sexbot researchers to work overtime in the development of newer and better technologies that can bring enhanced experiences. People will want better robot sex, and even better robot sex, and better still robot sex, their sexual appetites becoming voracious as the technologies improve, bringing even higher levels of joy with each experience. And it is quite possible that the terms "sex maniac" and "nymphomaniac" will take on new meanings, or at least new dimensions, as what are perceived to be natural levels of human sexual desire change to conform to what is newly available—great sex on tap for everyone, 24/7.

Notes

INTRODUCTION

1. Libin, Alexander, and Elena Libin. "Person-Robot Interactions from the Robopsychologists' Point of View: The Robotic Psychology and Robotherapy Approach." *Proceedings of the IEEE* 92, no. 11 (November 2004): 1789–1803.
2. Riskin, Jessica. "Eighteenth-Century Wetware." *Representations,* no. 83 (2003): 97–125.
3. Riskin, Jessica. "The Defecating Duck or the Ambiguous Origins of Artificial Life." *Critical Enquiry* 29, no. 4 (2003): 599–633.
4. Kanda, Takayuki, Takayuki Hirano, Daniel Eaton, and Hiroshi Ishiguro. "Interactive Robots as Social Partners and Peer Tutors for Children: A Field Trial." *Human Computer Interaction* 19, no. 4 (2004): 61–84.
5. Walker, Beverly, Jocelyn Harper, Christopher Lloyd, and Peter Caputi. "Methodologies for the Exploration of Computer and Technology Transference." *Computers in Human Behavior* 19, no. 5 (September 2003): 523–35.
6. Turkle, Sherry, Will Taggart, Cory Kidd, and Olivia Daste. "Relational Artifacts with Children and Elders: The Complexities of Cybercompanionship." *Connection Science* 18, no. 4 (December 2006): 347–61.
7. Kanda, Takayuki, Hiroshi Ishiguro, Michita Imai, and Tetsuo Ono. "Development and Evaluation of Interactive Humanoid Robots." *Proceedings of the IEEE* 92, no. 11 (November 2004): 1839–50.
8. Ibid.
9. Ibid.

CHAPTER 1

1. Fraley, Chris, and Philip Shaver. "Adult Romantic Attachment: Theoretical Developments, Emerging Controversies, and Unanswered Questions." *Review of General Psychology* 4 (2000): 132–54.

2. Kleine, Susan, and Stacey Baker. "An Integrative Review of Material Possession Attachment." *Academy of Marketing Science Review* (online publication) 8, no. 1 (2004). Available at www.amsreview.org/articles/kleine01-2004.pdf.

3. Belk, Russell. "Possessions and the Extended Self." *Journal of Consumer Research* 15 (September 1988): 139–68.

4. Grayson, Kent, and David Shulman. "Indexicality and the Verification Function of Irreplaceable Possessions: A Semiotic Analysis." *Journal of Consumer Research* 27 (June 2000): 17–30.

5. Pines, Ayala. *Falling in Love: Why We Choose the Lovers We Choose.* New York: Routledge, 1999.

6. Byrne, Donn, and Sarah Murnen. "Maintaining Loving Relationships." In *The Psychology of Love,* ed. Robert Sternberg and Michael Barnes. New Haven, CT: Yale University Press, 1998, 293–310.

7. Pines, Ayala. *Falling in Love.*

8. Livine, Deb. "Virtual Attraction: What Rocks Your Boat." *Cyberpsychology and Behavior* 3, no. 4 (August 2000): 565–73.

CHAPTER 2

1. Australian Broadcasting Corporation. *Animal Attraction.* TV program, 2001.

2. Rynearson, Edward. "Humans and Pets and Attachment." *British Journal of Psychiatry* 133 (1978): 550–55.

3. Wilson, Edward O. *Biophilia: The Human Bond with Other Species.* Cambridge, MA: Harvard University Press, 1984.

4. Kellert, Stephen. "The Biological Basis for Human Values of Nature." In *The Biophilia Hypothesis,* Keller, Stephen, and Edward O. Wilson, Washington, D.C.: Island Press, 1993, 42–69.

5. Headey, Bruce. "Pet-ownership: Good for Health?" *Medical Journal of Australia* 179 (November 3, 2003): 460–61.

6. Melson, Gail. *Why the Wild Things Are: Animals in the Lives of Children.* Cambridge, MA: Harvard University Press, 2001.

7. Robin, Michael, and Robin ten Bensel. "Pets and Socialisation of Children." *Marriage and Family Review* 8 (1985): 63–78.

8. Lewis, Thomas, Fari Amini, and Richard Lannon. *A General Theory of Love.* New York: Vintage Books, 2001.

9. Turkle, Sherry. *The Second Self: Computers and the Human Spirit.* New York: Simon and Schuster, 1984.

CHAPTER 3

1. Turkle, Sherry. *The Second Self: Computers and the Human Spirit.* New York: Simon and Schuster, 1984.

2. Young, Robert. *Mental Space.* London: Process Press, 1994.

3. Young, Robert. "Transitional Phenomena: Production and Consumption." In *Crises of the Self: Further Essays on Psychoanalysis and Politics,* ed. Barry Richards. London: Free Association Books, 1989, 56–72.

4. Turkle, Sherry. "Diary." *London Review of Books* (April 20, 2006): 36–37.
5. Turkle, Sherry. *The Second Self: Computers and the Human Spirit.*
6. Holland, Norman. "The Internet Regression." 1996. Available at www.rider.edu/ ~suler/psycyber/holland.html.
7. Weizenbaum, Joseph. *Computer Power and Human Reason: From Judgment to Calculation.* New York: W. H. Freeman, 1976.
8. Nass, Clifford, and Youngme Moon. "Machines and Mindlessness: Social Responses to Computers." *Journal of Social Issues* 56, no. 1 (2000): 81–103.
9. Moon, Youngme. "Intimate Exchanges: Using Computers to Elicit Self-Disclosure from Consumers." *Journal of Consumer Research* 26, no. 4 (March 2000): 323–39.
10. Ibid.
11. Nass, Clifford, and Youngme Moon. "Machines and Mindlessness: Social Responses to Computers."
12. Turkle, Sherry. "Diary."
13. Koller, Marvin. *Families: A Multigenerational Approach.* New York: McGraw-Hill, 1974.
14. Turkle, Sherry. "New Media and Children's Identity" (edited summary). Media In Transition Conference, MIT, October 1999. Available at web.mit.edu/m-i-t/ conferences/m-i-t/summaries/plenary1_st.html.
15. Turkle, Sherry. "Diary."
16. Ibid.
17. Beck, Alan, Nancy Edwards, Batya Friedman, and Peter Khan. "Value Sensitive Design: Robotic Pets and the Elderly." Available at www.ischool.washington .edu/vsd/projects/aiboelderly.html.
18. Turkle, Sherry. "Diary."

CHAPTER 4

1. Breazeal, Cynthia. *Designing Sociable Robots.* Cambridge, MA: MIT Press, 2002.
2. McGill, Michael. *The McGill Report on Male Intimacy.* New York: Holt, Rinehart and Winston, 1985.
3. Brave, Scott, Clifford Nass, and Kevin Hutchinson. "Computers That Care: Investigating the Effects of Orientation of Emotion Exhibited by an Embodied Computer Agent." *International Journal of Human-Computer Studies* 62 (2005): 161–78.
4. Turkle, Sherry. *The Second Self: Computers and the Human Spirit.* New York: Simon and Schuster, 1984.
5. Libin, Alexander, and Elena Libin. "Person-Robot Interactions from the Robopsychologists' Point of View: The Robotic Psychology and Robotherapy Approach." *Proceedings of the IEEE* 92, no. 11 (November 2004): 1789–1803.
6. Levine, Deb. "Virtual Attraction: What Rocks Your Boat." *Cyberpsychology and Behavior* 3, no. 4 (August 2000): 565–73.
7. Kim, Jong-Hwa, Kang-Hee Lee, Yong-Duk Kim, Bum-Joo Lee, and Jeong-Ki Yoo. "The Origin of Artificial Species: Humanoid Robot HanSaRam." Proceed-

ings 2nd International Conference on Humanoid, Nanotechnology, Information Technology, Communication and Control, Environment and Management, Manila, 2005.

8. Hodder, Harbour. "The Future of Marriage." *Harvard Magazine,* November–December 2004, 38–45.

9. Cott, Nancy. *Public Vows: A History of Marriage and the Nation."* Cambridge, MA: Harvard University Press, 2000.

10. Ibid.

11. Hodder, Harbour. "The Future of Marriage."

12. Hatfield, Elaine, and Susan Sprecher. "Men's and Women's Preferences in Marital Partners in the United States, Russia and Japan." *Journal of Cross-Cultural Psychology* 26, no. 6 (November 1995): 728–50.

13. Søby, Morten. "Collective Intelligence—Becoming Virtual." Paper 16, Kongress der Deutschen Gesellschaft für Erziehungswissenschaft: Medie-Generation, University of Hamburg, 1998.

14. Virilio, Paul. *The Art of the Motor.* Minneapolis: University Press of Minnesota, 1995.

15. Haraway, Donna. "A Manifesto for Cyborgs." *Socialist Review* 80 (1985): 65–108. Reprinted with minor revisions and corrections to notes, in *Coming to Terms: Feminism, Theory, Politics,* ed. Elizabeth Weed. New York: Routledge, 1989, 173–204.

16. Author unknown. "Life with a Robot. Wakumaru." Available at http://www.mhi.co.jp/kobe/wakamaru/english/about/index.html.

17. Levine, Deb. "Virtual Attraction: What Rocks Your Boat."

INTRODUCTION TO PART TWO

1. Bloch, Iwan. *The Sexual Life of Our Time,* trans. M. Eden Paul. London: Rebman, 1909.

2. Schwaeblé, René. *Les Détraqués de Paris: Études documentaries.* Paris, 1905.

3. Bloch, Iwan. *The Sexual Life of Our Time.*

CHAPTER 5

1. Suler, John. "Mom, Dad, Computer (Transference Reactions to Computers)," revised version 1998 (originally posted 1996). Available at http://www.rider.edu/~suler/psycyber/computertransf.html.

2. Holland, Norman. "The Internet Regression." Norman Holland, 1996. Available at www.rider.edu/~suler/psycyber/holland.html.

CHAPTER 6

1. Loebner, Hugh. "Being a John." In *Prostitution: On Whores, Hustlers, and Johns,* ed. James Elias, Vern Bullough, Veronica Elias, and Gwen Brewer. Amherst, NY: Prometheus Books, 1998, 221–25.

2. International Women's Rights Action Watch. *Country Report: The Philippines,* 2003. Available at http://iwraw.igc.org/publications/countries/philippines.htm.

3. Overall, Christine. "What's Wrong with Prostitution? Evaluating Sex Work." *Signs* (Summer 1992): 705–21.
4. Sánchez Taylor, Jacqueline. "Dollars Are a Girl's Best Friend? Female Tourists' Sexual Behavior in the Caribbean." *Sociology* 35, no. 3 (2001): 749–64.
5. Bowen, Nigel. "Sugar Mamas." *Bulletin,* Sydney, June 9, 2004.
6. Sánchez Taylor, Jacqueline, and Julia O'Connell Davidson. "Fantasy Islands: Exploring the Demand for Sex Tourism." In *Sun, Sex and Gold: Tourism and Sex Work in the Caribbean,* ed. Kampadoo Kamala. New York: Rowman and Littlefield, 1999.
7. Sánchez Taylor, Jacqueline, and Julia O'Connell Davidson. "Travel and Taboo: Heterosexual Sex Tourism to the Caribbean." In *Regulating Sex: The Politics of Intimacy and Identity,* ed. Laurie Schaffner and Elizabeth Bernstein. London: Routledge, 2005, 83–100.
8. Plumridge, Elizabeth, Jane Chetwynd, Anna Reed, and Sandra Gifford. "Discourses of Emotionality in Commercial Sex: The Missing Client Voice." *Feminism and Psychology* 7, no. 2 (1997): 165–81.
9. Bernstein, Elizabeth. "Desire, Demand, and the Commerce of Sex." In *Regulating Sex: The Politics of Intimacy and Identity,* ed. Laurie Schaffner and Elizabeth Bernstein. New York: Routledge, 2005, 101–28.
10. Holzman, Harold, and Sharon Pines. "Buying Sex: The Phenomenology of Being a John." *Deviant Behavior* 4 (1982): 89–116.
11. Kernsmith, Roger. "Social Bonds as Motivators for Deviant Behavior: An Unobtrusive Analysis of Prostitute Clients." Eastern Michigan University, 2001.
12. Winick, Charles. "Prostitutes' Clients Perceptions of the Prostitutes and of Themselves." *International Journal of Social Psychiatry* 8, no. 4 (1992): 289–99.
13. de Hemptinne, Gerald. "Hookers Learn to Give 'More than Sex.'" Available at http://iafrica.com/loveandsex/features/232211.htm.
14. Monto, Martin. "Prostitution and Fellatio." *Journal of Sex Research* 58, no. 2 (May 2001): 140–45.
15. Ibid.
16. Monto, Martin. Personal communication.
17. Bernstein, Elizabeth. "Desire, Demand, and the Commerce of Sex."
18. McKeganey, Neil, and Marina Barnard. *Sex Work on the Streets: Prostitutes and Their Clients.* Buckingham, UK: Open University Press, 1996.
19. Holzman, Harold, and Sharon Pines. "Buying Sex: The Phenomenology of Being a John."
20. Monto, Martin. "Prostitution and Fellatio."
21. Plumridge, Elizabeth, Jane Chetwynd, Anna Reed, and Sandra Gifford. "Discourses of Emotionality in Commercial Sex: The Missing Client Voice."
22. Laner, Mary. "Prostitution as an Illegal Vocation: A Sociological Overview." In *Deviant Behavior: Occupational and Organizational Bases,* ed. Clifton Bryant. Chicago: Rand-McNally, 1974, 406–18.
23. Lyman, Randy. "The Ethics of Sex Therapy," 2001. Available at www.disbled.gr/at/?p=1650.
24. Blanchard, Vena. "The Tapestry of Surrogate Partner Therapy." The School of ICASA Web site, 2001. Available at www.icasa.co.uk.

25. Noonan, Raymond. "Sex Surrogates: The Continuing Controversy." In *Continuum Complete Encyclopedia of Sexuality*, ed. Nell Noonan and Robert Francoeur. New York: Continuum, 2003, 1279–1281.
26. Roberts, Barbara. "The Sexual Surrogate," *InnerSelf*, 2000. Online magazine available at http://innerself.com/Sex_Talk/Sexual_Surrogate.htm.
27. Ibid.

CHAPTER 7

1. Maines, Rachel. *The Technology of Orgasm*. Baltimore: Johns Hopkins University Press, 1999.
2. Granville, Joseph. *Nerve-Vibration and Excitation*. London: J. & A. Churchill, 1883.
3. Ibid.
4. Maines, Rachel. "Declaration Under Penalty of Perjury." Affidavit for the United District Court, Northern District, Alabama, April 30, 2001.
5. United States District Court Northern District of Alabama Northeastern Division, Civil Action No. CV-98-S-1938-NE, 2002.
6. Rubenstein, Steve. "Texas Housewife Busted for Hawking Erotic Toys." *San Francisco Chronicle*, December 16, 2003.
7. Davis, Clive, Joani Blank, Hung-Yu Lin, and Consuelo Bonillas. "Characteristics of Vibrator Use Among Women." *Journal of Sex Research* 33, no. 4 (1996): 313–20.
8. Heslinga, Klaas, Antonius Schellen, and Arie Verkuyl. *Not Made of Stone: The Sexual Problems of Handicapped People*. Leyden, Netherlands: Stafleu's Scientific Publishing Company, 1974.
9. Robert Trost, personal communication.
10. Tabori, Paul. *The Humor and Technology of Sex*. New York: Julian Press, 1969.
11. Author unknown. "Alma's Life," 2005. Available at http://www.alma-mahler.at/engl/almas_life/puppet3.html.
12. Lamarr, Hedy. *Ecstasy and Me: My Life as a Woman*. New York: Fawcett World Library, 1966.
13. Laslocky, Meghan. "Real Dolls: Love in the Age of Silicone." Available at www.saltmag.net/images/pdfs/RealDollsPDF.pdf. (An earlier and shorter version of this article appeared on www.salon.com under the title "Just Like a Woman," October 11, 2005.)
14. Author unknown. "Silicone Ladies at Your Beck and Call." *Japan Today*, April 16, 2005.
15. Connell, Ryann. "Rent-a-Doll Blows Hooker Market Wide Open." *Mainichi Daily News*, December 16, 2004. Available at http://mdn.mainichi.co.jp/waiwai/0412/1216doll.html.
16. Tabori, Paul. *The Humor and Technology of Sex*.
17. Block, Susan (quoting Gordan, Mel). "Our Night of Weimar Love." Available at www.bigomagazine.com/features05/SBweimar.html.
18. Rheingold, Howard. *Virtual Reality*. London: Martin Secker and Warburg, 1991.
19. Ibid.

20. Balderston, Meredith, and Timothy Mitchell. "Virtual Vaginas and Pentium Penises: A Critical Study of Teledildonics and Digital S(t)imulation." Conference on Performance, Pedagogy, and Politics in Online Spaces, University of Maryland, April 2001. Available at www.georgetown.edu/users/baldersm/essays/teledildonics.pdf.
21. Maheu, Marlene. "The Future of Cyber-Sex and Relationship Fidelity," *Brave New World* booklet, 1999, available at www.selfhelpmagazine.com/booklet/sexuality.html.

CHAPTER 8

1. Kinsey, Alfred, Wardell Pomeroy, and Clyde Martin. *Sexual Behavior in the Human Male.* Philadelphia: B. Saunders Company, 1948.
2. Taylor, Gordon. *Sex in History.* London: Thames and Hudson, 1953.
3. Tannahill, Reay. *Sex in History.* London: Sphere Books (Cardinal), 1989.
4. Kon, Igor. "Russia." In *The Continuum Complete Encyclopedia of Sexuality,* 2004, ed. Robert Francoeur and Raymond Noonan. Available at http://www.kinseyinstitute.org/ccies/ru.php.
5. Taylor, Gordon. *Sex in History.*
6. Ibid.
7. Plant, David. "The Rump Parliament." British Civil Wars and Commonwealth Web site at http://www.british-civil-wars.co.uk/glossary/rump-parliament.htm.
8. Juhasz, Anne. "Variations in Sexual Behavior." *Journal of Research and Development in Education* 16, no. 2 (1983): 53–59.
9. Taylor, Gordon. *Sex in History.*
10. Michael, Robert, John Gagnon, Edward Laumann, and Gina Kolata. *Sex in America: A Definitive Survey.* Boston: Little, Brown & Company, 1994.
11. Stengers, Jean, and Anne van Neck. *Masturbation: The History of a Great Terror.* New York: Palgrave/St. Martin's, 2001.
12. Pomeroy, Wardell. *Boys and Sex.* New York: Delacorte Press, 1968.
13. Morris, Maxwell. "The Three R's of Sex." *Journal of Religion and Health* 17, (1978): 48–56.
14. Baumeister, Roy, Kathleen Catanese, and Kathleen Vohs. "Is There a Gender Difference in Strength of Sex Drive? Theoretical Views, Conceptual Distinctions, and a Review of Relevant Evidence." *Personality and Social Psychology Review* 5, no. 3 (2001): 242–73.
15. Whitehouse, David. "Japanese Develop 'Female' Android." BBC News Web site, July 27, 2005.
16. Snell, Joe. "Impacts of Robotic Sex." *The Futurist: A Magazine of Forecasts, Trends, and Ideas about the Future* (July–August 1997): 32–34.

Index

9/09
1/0
8/08